3

PC

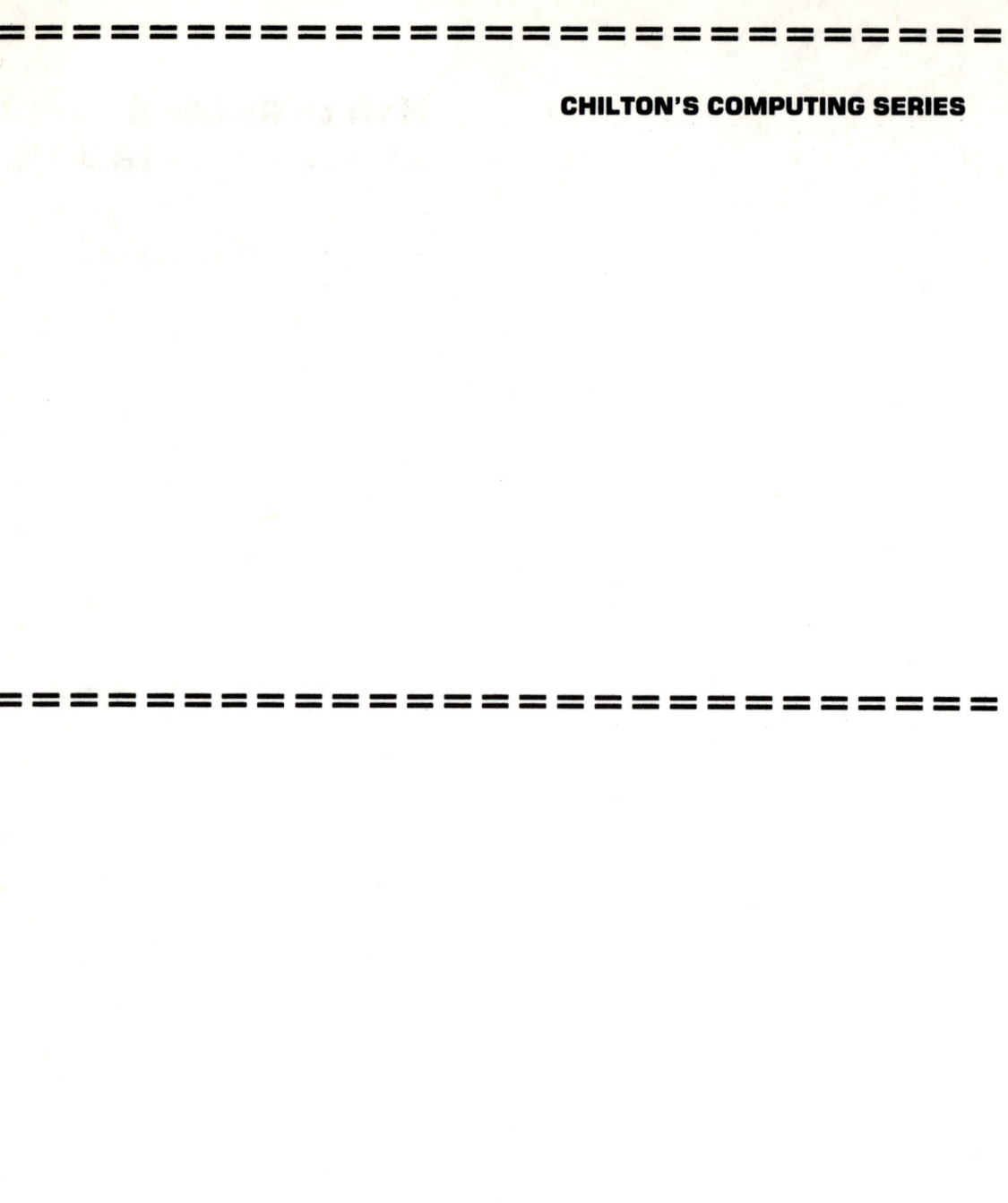

CHILTON'S COMPUTING SERIES

How to Repair & Maintain Your IBM PC

Gene B. Williams

CHILTON BOOK COMPANY

Radnor, Pennsylvania

To the two people most responsible
for this book, my parents.
Also to my wife, Cindy,
who only rarely interrupted with,
"Don't you think it's about time to take
a break from that computer?"

Copyright © 1984 by Gene B. Williams
All rights reserved
Published in Radnor, Pennsylvania 19089 by Chilton Book Company
Designed by Arlene Putterman
Line drawings by Adrian Ornik
Manufactured in the United States of America

Library of Congress Cataloging in Publication Data
Williams, Gene B.
  How to Repair and maintain your IBM PC.
  (Chilton's computing series)
  Includes Index.
    1. Microcomputers—Maintenance and repair.
2. IBM Personal Computer. I. Title. II. Series.
TK7887.W55   1984   621.3819'584   84-45156
ISBN 0-8019-7537-9

Chilton's Computing Series

1 2 3 4 5 6 7 8 9 0   0 9 8 7 6 5 4 3 2 1

Contents

Williams: Repair & Maintenance of Your IBM PC (Chilton)

Williams: Repair & Maintenance of Your IBM PC (Chilton)

====================================

Preface

A few years ago I bought my first computer. I was facing a tight deadline. The computer was checked out before I packed it into my car and brought it to the office. Back in the office, it was plugged in, booted up with the program—nothing! The screen was blank.

I called the seller to explain the problem. He said that he couldn't be out until the following Wednesday. So much for my deadline. The only solution was to try a diagnosis by phone.

He told me to open the case. I was convinced that the second I did that, the machine would be ruined for all time. A look inside confirmed my fears. There were what seemed to be several hundred memory boards along with enough mysterious components to launch the next deep space probe. I had about as much desire to touch the insides of the machine as I have to jump out of a moving car.

Under the dealer's direction each board was pulled, cleaned, and inserted back into the mother board. What should have been a 30-second job took me nearly 30 minutes. I was so sure that every movement would break something important that I did everything *very* slowly.

I tried to get the computer to operate again, thinking that a reset might cure the trouble. Absolutely nothing! The dealer repeated, "I really can't make it there before Wednesday. Oh, by the way, have you checked the contrast control? Maybe you bumped it."

"Contrast control? What contrast control?"

"It's on the left, beneath the keyboard."

I reached under and felt a little wheel. "You mean that little wheel thingy?"

Suddenly the screen came to life with all the signs and symbols it was supposed to display. The problem was solved and I met the deadline.

As the months went by I ran into other problems. Each time I went through the same feelings of helplessness. A computer is such a complicated piece of machinery—isn't it? Doesn't it require years of training and experience, plus a room full of special tools and test equipment, to repair?

Most of my problems turned out to be minor. Repair was usually handled with tools no more complex than a screwdriver. A few times I had to get "technical" and use an ordinary voltmeter.

Then I purchased an IBM PC. The knowledge I'd gained with my first computer carried over. From conversations with other computer owners, it became obvious that most not only knew very little about their machines, but were on the verge of being terrified of them.

As a result, this handbook and guide was developed. It will help you in taking care of the most common failures in a personal computer system. With rare exception, the average person *can* do it. A computer is nothing all that grand and mysterious. It's just a machine. Sophisticated perhaps, but a machine all the same. There are only so many reasons why a machine fails to work.

If you're standing in a book store reading this, the purpose of this book is to save you *at least* ten times the cost of the book. More important to some, it will save you a great deal of time, both in waiting for the technician and in driving to and from the shop. BUY IT! If you've already bought this book, congratulations! You won't regret it.

==============================

How to Repair & Maintain Your IBM PC

====================================

Introduction

No other brand of personal home computer has had the impact of the IBM PC. Other home computers came first, but these took an instantaneous back seat when "Big Blue" announced the PC. Since its introduction, dealers have had difficulty in keeping up with the demand. To help meet the demand many companies began to make PC look-alikes.

The popularity of home computers is due largely to the fact that they are becoming easier and easier to operate. Not many years ago the person who had a computer in his home was thought of as being a genius (or a nut). There were very few commercial programs available, which meant that the owner of a home computer had to have a solid knowledge of programming.

Ask any ten people today. Several will own a computer. More yet will be in the category of "We're thinking about it." Nine out of ten are likely to have some kind of computer around—a pocket calculator if nothing else.

Almost anyone can operate a computer. You might enjoy your computer more if you learn some programming, but all you really need are preprogrammed diskettes. Push one in, press a few buttons, and you're ready to go.

Meanwhile, the cost for technical work—repair, installing add-ons, etc.—has jumped. Charges of $60 per hour or more for labor alone are common. Some shops charge a $150 minimum to just *look* at a malfunctioning machine. Average down time for such a repair is three days. Having the machine tied up for a week or more isn't unusual. Should the computer owner need the technician to come to his home or place of business, all costs increase.

To avoid these high costs, some owners purchase a repair contract. Normal cost is between 5% and 10% of the purchase price per year. A fee of 20% isn't unheard of. Several sources list a cost of 1% per month as the minimum repair cost an owner should expect *without* such a contract.

To make things even more depressing, it has been said that 95% of all repairs and other technical work could be taken care of by the computer owner *without special tools or technical background.* (See "Tools You'll Need" at the end of this Introduction.) There is usually no need for you to spend hundreds of dollars or to wait a week (or even a day) while the repair is being done, nor even waste the time driving to and from the shop.

My collaborator on this book is Bill Belisle, service manager for Computerland of Mesa, a computer store in Mesa, Arizona. He has repaired computers of all kinds. When he and I first met to discuss this book he said, "About two-thirds of all the repairs I do require nothing more complicated than my fingers." My own experience verifies this statement.

ADVANTAGES OF THIS BOOK

The purpose of this book is to show you just how easy it is to diagnose and repair most malfunctions. It will also show you how easy it is to maintain a system to reduce repairs. You don't need a knowledge of electronics. It helps, but you can handle many normal repairs without it.

Chapter 1 will acquaint you with the "rules of the game." Its purpose is to show you what can cause trouble or damage, both to you and to the computer. Dangerous spots are identified to prevent your getting a shock. Precautions are given so you can avoid making costly mistakes. Read this chapter thoroughly, put the information to use, and you are highly unlikely to run into trouble while working on your computer.

Chapter 2 will show you how to diagnose malfunctions and how to get your IBM to diagnose itself. The people at IBM designed the system so that it can tell you what's wrong, where it's wrong, and even how to fix it. (Very clever, these computers.) Additional tips are provided in this chapter on how to track down a problem. Chapters 3, 4, 5, and 6 take you deeper into the specific problem areas.

Proper maintenance can reduce repair costs dramatically. Chapter 7 tells you how to reduce problems and repair costs by prevention. After you've read this chapter you'll know what to do and what *not* to do. Knowing how to do many of the little tasks to keep your IBM happy and healthy can save you a great deal of time and expense (and frustration).

Unless you knew ahead of time exactly what you needed in a computer system (in which case you're a rare individual), there will come a time when you want to add something to your computer. It might be a second printer, a phone modem, or some additional memory. Whatever you care to add, you'll find the help you need in Chapter 8.

By following the steps in Chapter 2, you'll have a good idea as to what the malfunction is. For those jobs you don't want to handle yourself (or can't), you'll begin with a good idea of what has gone wrong. This in turn will help reduce repair costs and the risk of being ripped off by unnecessary repairs. Chapter 9 gives you some additional tips to help you find a reliable technician and how to deal with this person.

Chapter 10 is a troubleshooting guide listing "symptoms" and "cures," along with references to the appropriate repair chapter. At the back of the book are some tables and charts that will be useful to the home technician and a glossary where you can look up unfamiliar terms.

Even if you have no background in electronics, you can still handle most repairs and maintenance. However, the more you know, the easier it will be. Your local library or bookstore will most likely have some books on basic and advanced electronics. The less you know, the more important it is that you pick up such a book and learn a bit about electronics. Your goal isn't to become an electronics whiz; you just want a more thorough understanding of what is going on and, consequently, what is going wrong and why.

It is suggested that you read through this entire book before tearing into your computer. Even those parts you don't think you'll need are important. This will help give you an overall idea of what your computer does and how it does it. Don't be in so much of a hurry that you end up causing more problems than you started with.

Before you start yanking out parts or devices, go through Chapter 2 (Diag-

---

THE FIRST STEPS

1. Read the book thoroughly
2. Read Chapter 1 again (for safety tips)
3. Perform diagnosis (Chapter 3)
4. Read applicable repair chapter (3–6)
5. Repair or replacement
6. Consult a professional if needed (Chapter 9)
7. Back in operation again!

---

nosis) carefully. If you just tear into the machine without careful diagnosis of the problem, you could spend hours trying to find out what is wrong (if you ever find the problem at all). Proper diagnosis will guide you to the source of the malfunction. If the problem is with the drives, why spend time with the RAM memory? The few minutes you spend with Chapter 2 can save you hours of wasted effort.

Use this book correctly and it should save you at least ten times its cost for the first repair. Instead of spending that 1% per month ($600 per year on a $5000 system) or that 10% per year ($500 on the $5000 system), you could be spending only $50 per year, if that much. Why spend more than you have to?

Expenses aren't in money alone. There is the time involved. Time waiting for the technician to come, or time wasted in driving to and from the shop and waiting to have the technician look at your computer. This could amount to an entire day if you live a considerable distance from the shop. Then there is down time, when your valuable computer is useless.

You bought your computer to save time. Why use up what you've saved in waiting hours or days only to find out that you could have fixed the trouble yourself in a few seconds?

An added benefit will be that you will understand your system better. You'll know what can go wrong. You'll learn how to fix the most common problems, and even how to prevent troubles.

TOOLS YOU'LL NEED

The IBM PC is a highly sophisticated machine. It is also extremely well designed. The more you work with it, the more obvious this will become. Because of this intelligent design, you can troubleshoot and repair your PC with nothing more than a few basic tools. You bought several of them when you bought your computer. The main tool is the computer itself. The DOS and diagnostics diskettes are both powerful tools for repair and maintenance. With this book as a guide, and using your own intelligence, you have just about everything you need.

Unlike so many machines of "modern" manufacture, nothing in the PC requires a special or expensive tool. Chances are, you already have all the tools you need. If you don't, the cash outlay to equip yourself will be small.

1. *Screwdrivers.* Using nothing more than a standard screwdriver and a small-headed Phillips (with insulated handles to protect yourself), you can just about completely disassemble your PC. The cabinet is attached with screws; the boards (including the main system board) are held with screws; the drives are held in place by screws, and are held together by screws. There is very little in the PC that does not come apart with the twist of an ordinary screwdriver.

Williams: Repair & Maintenance of Your IBM PC (Chilton)

FIG. I—1 The three most powerful tools—the IBM PC and the DOS and diagnostics diskettes.

2. Multimeter. To test for voltages, to measure component values, and to check for continuity, you will need a fair-quality multimeter. It doesn't have to be a fancy digital multimeter. Voltages measured will be in the 5 and 12 volt dc range, and 120 ac. It should also be capable of measuring resistance (in ohms). Accuracy *is* important, especially when measuring voltages. If you're not familiar with the use of a multimeter, practice using it before probing inside the computer. For example, take readings of the various outlets in your home to check for ac voltage. Use the meter to check some batteries (dc voltage). If you have some old resistors lying around, check these for correct resistance. It doesn't take long to learn how to use it efficiently and accurately.

3. Digital soldering tool. If you intend to change single components, you will have to have a high-quality soldering gun, and one designed specifically for digital circuits. If you will not be replacing components, you won't need this tool, and you can put off the investment (about $50) until you need it. The tool should have a rating of no higher than 40 watts. If possible, the tip should be grounded to avoid electrical damage to delicate components.

4. IC extractor. Many of the ICs (integrated circuits) in your computer are plugged into sockets. Replacement of these is a simple job, but it is a risky one if all you use is your fingers because the many prongs of the IC are easily damaged. The extractor is used to remove the chip from its socket safely. This is when most of the damage to the prongs occurs.

OPTIONAL TOOLS

With the two screwdrivers and the multimeter, you'll be able to take care of almost any problem and any add-on. The soldering gun and IC extractor will be used only if you intend to handle such detailed repairs or additions. Other tools, such as those listed below, will merely make the job easier.

1. *Needlenose pliers.* You'll rarely need regular pliers. However, having a needlenose pliers at hand makes retrieving dropped parts easier. They also help to remove parts that are being desoldered.

2. *Nut drivers.* These are like fixed socket wrenches. Many of the screws in the PC can be removed and installed with either the screwdriver or a hex nut driver. Quite often a nut driver makes removal and replacement of the screws easier, faster, and safer.

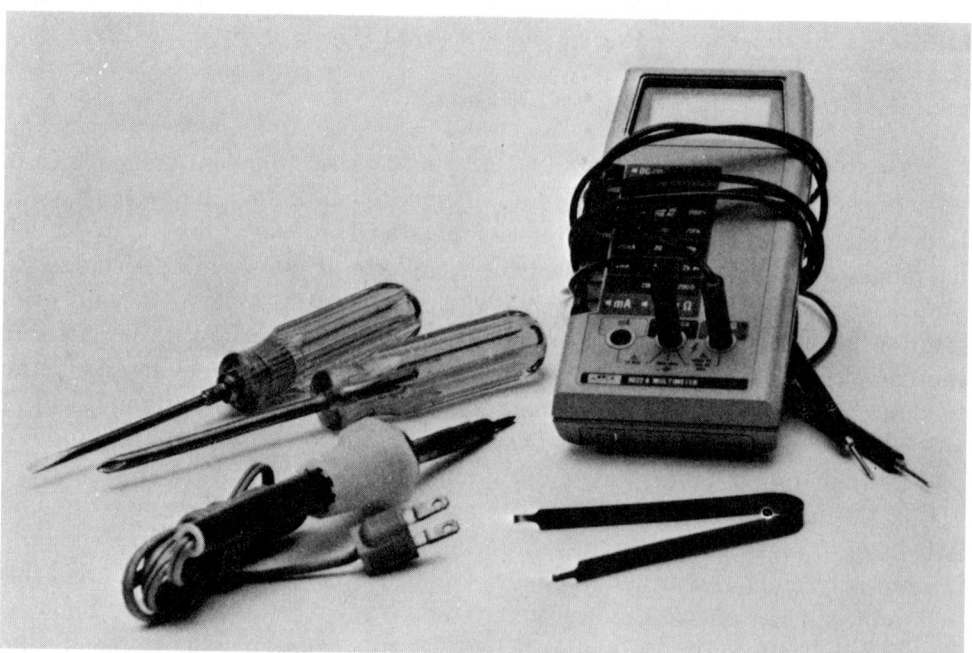

FIG. I–2 The tools you'll need: screwdrivers, multimeter, soldering tool, and IC extractor.

Williams: Repair & Maintenance of Your IBM PC (Chilton)

INTRODUCTION

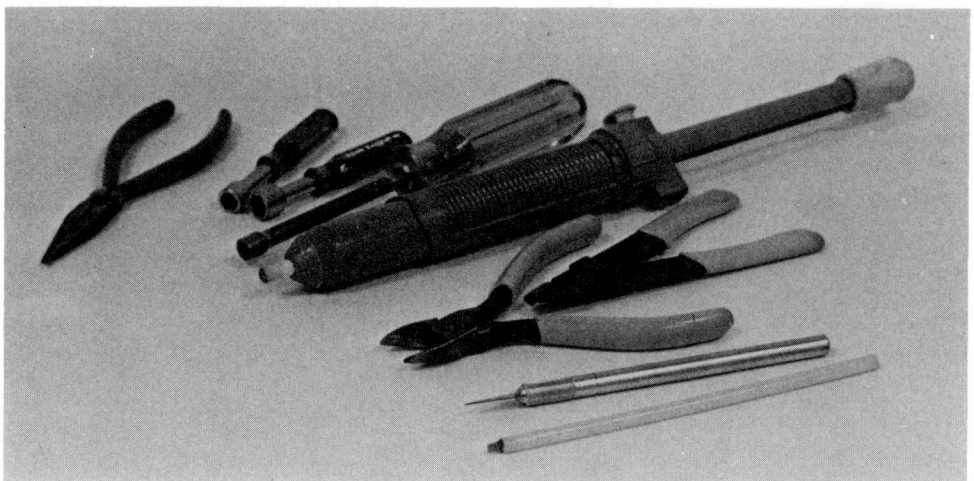

FIG. I–3 Tools to make the job easier: needlenose pliers, nut drivers, desoldering tool, wire cutters, and knife.

3. *Desoldering tool.* This is a fancy name for a heat-resistant syringe. Its function is to suck away solder from a heated joint. Without it, removing components is possible, but difficult.

4. *Wire cutter.* New components often have metal leads which are too long. This means that they must be clipped to the proper length. A wire clipper handles this job correctly and efficiently. Some pliers have built-in clippers. These are suitable for cutting wire but are not meant to trim the leads of components that are soldered in place.

5. *Knife.* A small sharp knife can be used for many jobs. Used correctly it can be a valuable tool.

==

Best Results/
Minimal Time
1

Have you ever watched a child take apart a toy? The usual way is for the pieces to go flying in all directions, without any order or planning. Parts that don't come off easily are broken off. Some roll under the couch; others get stepped on; some just disappear.

When it comes to reassembling the toy, the child rarely has the slightest idea of what to do or how to do it. That's when the child brings it to mommy or daddy with big, sad, wet eyes and says, "My toy broke. Fix it for me."

A fair number of computer repair jobs come as a direct result of an adult "child" getting inside his electronic "toy" to find out how it works, or in an attempt to repair some malfunction. Quite often what started out as a minor problem is klutzed into something expensive.

Sure, you can save yourself hundreds of dollars a year by doing your own repairs and maintenance and by installing any add-on equipment yourself. Approach it incorrectly, however, and it can end up costing you many times what the repair should have cost. Sometimes in ways you didn't expect.

YOUR SAFETY

Nothing is more important than your own safety. If you do something that destroys a circuit in the computer, that circuit can be replaced. Even if repair of the malfunction costs several thousand dollars, you can take out a loan if need be and get it paid off over a period of years. If you let something happen

Williams: Repair & Maintenance of Your IBM PC (Chilton)

BEST RESULTS/MINIMAL TIME

to you—well, there is no such thing as taking out a loan for more life, at any interest rate.

There are actually very few danger spots in your computer. Even while in operation the voltage inside the computer (past the power supply) is either 5 volts or 12 volts, both dc (direct current). The amount of current flowing is so tiny that you wouldn't even feel it.

The dc current used in the operation of most digital circuits isn't at all dangerous to a human being. However, there are certain places where the voltage and current aren't quite so safe. These spots are usually where ac (alternating current) power comes into the system. Touch one of these places and you're in for a bad time.

EFFECTS OF CURRENT

The line coming into the power supply of your computer, and into most peripherals, is 120 volts ac. The amperage can be as high as the physical limits of the wire and the circuit breaker or fuse. Usually this means that the line is 120 volts with a current of at least 15 amps steady, plus a surge limit in the hundreds of amps. This is enough power to melt a metal rod, and more than enough power to kill.

Tests were done by the U.S. Navy to learn the effects of alternating current with a frequency of 60 cycles per second (cps). (The measure of frequency is sometimes reported in hertz, with one hertz being the same as one cps.) It was found that a tiny trickle of just 1 milliamp (.001 amps, or one one-thousandth of an amp) would produce a shock that could be felt. A current of 10 milliamps (.01 amps, or one one-hundredth of an amp) would cause the muscles to become paralyzed, making it impossible for the person to let go of the source of the shock. In fact, the spasms caused by this amount of current can cause the victim to grip the source more tightly. At 100 milliamps (one tenth of an amp) the shock is usually fatal if it continues for more than a second.

As you can see, it doesn't take much current to represent a severe hazard. If you carelessly touch a hot spot you will become a part of the power circuit. For a short time (until the fuse or circuit breaker pops) the current flows unhampered. You risk having a surge of perhaps a hundred amps flow through your body, which is over 2000 times as much as is needed to be fatal.

DANGER SPOTS

Any place there is an ac current presents a risk to you. Most of these spots are obvious and are easy to avoid. The danger begins with the wall outlet (or the

MOST DANGEROUS SPOTS

1. Wall outlet
2. Power cord
3. Power switches

4. Monitor
5. Filter capacitor
6. Printer and mechanical parts

circuit box if you ever get to fooling around there). It moves up through the wires and into the power switches.

The wires come into the equipment and are connected in such a way as to make it very difficult to touch the contacts. The power supply is sealed and has a sticker on top, warning you to STAY OUT! The power supply is one of the only real electrical dangers to you inside the computer. It has ac coming into it and ac going out (to the monitor). There is rarely a need to open the PC power supply. Repair is generally handled by replacing the entire unit.

Inside the power supply there is another danger, namely the filter capacitor. It looks like a small can. The ac comes into the power supply where it is changed in value to the 5 volts and 12 volts needed. It is also changed from alternating current to direct current. The filter capacitor helps to smooth out the flow. To do this it stores up current as it comes in, and then lets it flow out again in a steady stream.

Even after the computer is shut off, and even with the power cord pulled from the outlet, this capacitor can have a hefty charge inside. Theoretically it should drain itself of all charge in a matter of seconds. Normally there is no danger. But if something should go wrong with the circuitry, you won't know it until you touch the capacitor contacts—at which point you'll find out all too quickly.

The PC and most peripherals have power going directly to the switches. It's a common misconception that a switch is safe when it is in the "off" position. It is *never* safe unless the power cord has been removed from the outlet. If you happen to touch the incoming contacts, it would be the same as if you grabbed the bare power lines.

With the power off, anything past the switch can usually be considered safe. (When dealing with power, though, you're safer unplugging the cord from the outlet.)

Just beyond the switch is a fuse. With the switch "off," there is no current flowing through the fuse or into the computer (assuming that the switch is functioning properly). This fuse is for protection of the circuits inside, not for your protection. If a short circuit occurs, the power supply will begin to draw large

Williams: Repair & Maintenance of Your IBM PC (Chilton)

amounts of current. In a very short time this increased flow will cause serious damage. It could also cause a fire. The fuse helps to prevent this from happening.

The fuse of the PC is rated at 2 amps. The power supply in a printer might have a fuse rated at 4 amps. This simply means that if the current reaches a level higher than this, the fuse wire will melt and current will not flow beyond the fuse. For a fraction of a second more current *can* flow, however. Worse, if you create a short circuit across the fuse, that fuse will do nothing at all. Your body, the screwdriver, etc., becomes the new fuse. Normally this means that you're grabbing a bare wire with 120 volts and temporarily unlimited current.

Power supplies in some PCs have a fuse that can be changed by the user; others do not. Unless you're experienced and know what you're doing, you should not venture into the power supply of your PC.

With the power switched to "off," you can safely change the fuse, the fuse holder, and any other power handling components inside the power supplies of the peripherals (your printer, for example). This again assumes that the switch is operating correctly and that the wires are all connected as they should be connected. (In one shipment of several thousand monitors, the wiring was found to be connected backwards!) Before you begin, take a moment to get out the voltmeter to measure if there is any voltage present.

The monitor is another source of high voltage. There is the 120 volts ac coming into it, which is a danger in itself. There is always the chance that you have one of those monitors which was wired improperly. But the danger doesn't stop here.

The monitor is a CRT (cathode ray tube) which works by throwing electrons at the phosphor-coated screen. This requires a considerable charge. The larger the monitor is, the larger will be the voltage required to form an image. Even a small black and white monitor may require a few thousand volts. Color monitors require still more, sometimes as high as 25,000 volts. The current (amperage) is low, but this doesn't make the monitor safe.

The monitor has yet another danger, one that has nothing to do with electricity. The screen tube has a vacuum inside and thin glass walls. Striking the tube can cause an implosion, which will cause very sharp slivers of glass to be thrown around.

There is rarely a need to open the PC power supply. Repair is generally handled by replacing the entire unit. If you have internal problems with the PC power supply, leave the repair to a professional.

Williams: Repair & Maintenance of Your IBM PC (Chilton)

MEASURING VOLTAGE

Although most equipment manufacturers do their best to reduce the risks of accidents, you still have to be careful. Remember the case when thousands of monitors were shipped with the wires connected in the wrong order. A "hot" (active wire) was connected to a spot that should have been "dead." A technician who assumed that everything was as it should be could be in for a terrible shock—literally.

If you are going to be working on the main lines coming into a piece of equipment, the power switches, or anywhere that an ac current might be present, don't take the chance that the spot or contact is dead just because it is supposed to be. Measure the voltage. Better yet, unplug the equipment and then measure. In short,

1. Shut off power
2. Unplug
3. Test for current

Don't trust anything unless you've measured it and know that no current is

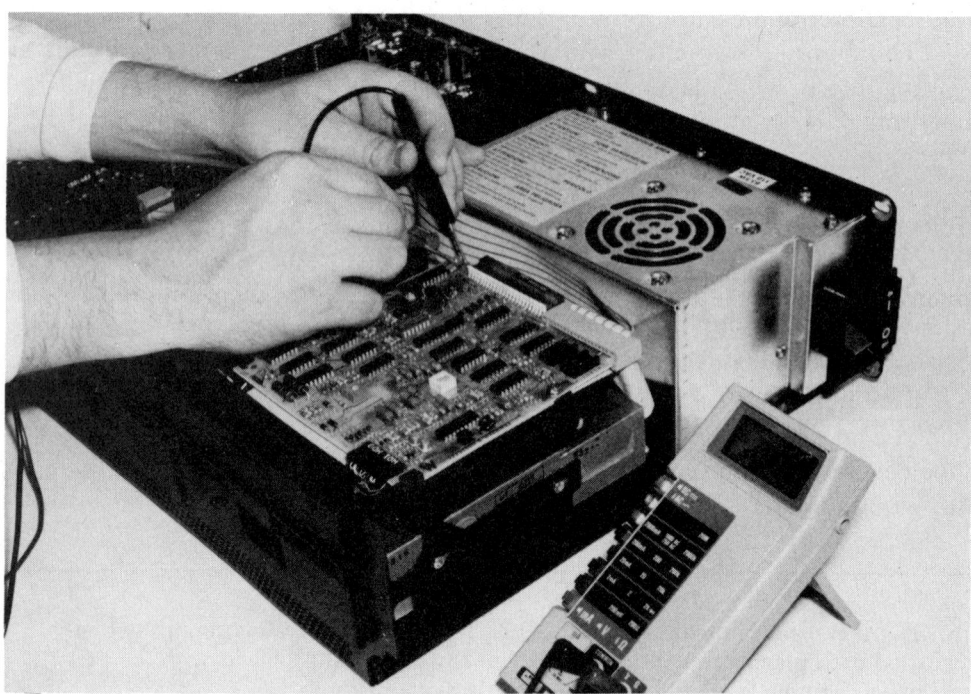

FIG. 1–1   Learn how to use a voltmeter.

Williams: Repair & Maintenance of Your IBM PC (Chilton)

BEST RESULTS/MINIMAL TIME

present. Even then, be careful. This rule applies whenever you're working around something electrical.

Testing for "hot" is easy to learn if you don't already know how. With a voltmeter or other testing device, touch one probe (usually black) to a known ground, such as the metal chassis of the computer. Be sure to hold the probe *only* by the insulated handle. The other probe (usually red) gets touched to the suspected point. Assuming that the meter is functioning properly, and that you've put it into the proper testing range, it will tell you if a charge is present and how large that charge is.

Setting the meter to the correct range is important. If you intend to measure the voltage at the monitor screen, don't have the meter set for 3 volts. (The setting should be in the thousands of volts. See the manual that came with your monitor.) If you're testing for ac, don't have the meter adjusted to test for dc. You're asking for trouble if you just jam the probes inside the computer without first looking to see if the meter is properly set.

RULES OF SAFETY

Working around electricity demands a set of safety rules. The first step is to shut off the power. The main switch will cut the flow to parts farther in. It doesn't protect you completely, though, since the wires between the switch and the wall outlet are still hot (active).

You might think that the best way to protect yourself completely would be to unplug the computer. This is true, with reservations. While pulling the plug does remove the current coming in from the wall outlet, it also removes the safety of a ground wire. This step is more to prevent damage to the computer than to yourself. Whether you unplug the computer or not depends on the circumstances.

With or without the plug, assume that all circuits are live and carry a potentially dangerous current. (The vast majority do not, but if you treat them as if they do you are unlikely to cause damage to either the computer or to yourself.)

You can't see electricity, nor can you tell by sight if a circuit is hot. The only immediate indication of power flowing inside the computer is the soft hum of the fan, which is easy to ignore once you become accustomed to it.

By making voltage measurements you assure that you're not sticking your fingers in places where damage will be done—damage to you or to the computer.

Never probe or poke inside the computer with any part of your body touching a conductive surface. Avoid leaning on the chassis, a metal work bench, or anything made of metal if you are reaching inside. Also take care that your feet aren't touching anything conductive (which includes a damp floor). In short, insulate yourself from your surroundings and from the equipment.

To further protect yourself use the "one-hand rule." This means, simply, don't reach in with both hands at the same time. Often the rule says that one hand should be kept in a pocket to avoid the temptation of breaking the rule.

The idea behind this rule is to prevent your body from becoming a part of a circuit. If just one hand touches a spot and your body is insulated from all conductive surroundings, the current has nowhere to go. If a second hand touches a hot spot, the current can enter the one hand, pass through your body and out through the other hand.

Any and all tools should have insulated handles. Touch the tools only by these handles. It's sometimes tempting to grab a part of the blade of a screwdriver for better control, for example. Don't do it! The insulation is on the handle for a reason.

Many people realize that grabbing a tool by the metal is foolish, and then forget that the jewelry or wristwatch they are wearing is made of metal. It will conduct current just as well as the shaft of a screwdriver—better if it's made of gold or silver. Necklaces are especially hazardous because they can dangle into dangerous spots, but the same caution applies to other jewelry, such as rings. Any piece of metal can touch a hot spot and carry the current into your body.

A dangling necklace or tie might also become entangled in some mechanical part. Inside the computer there are only two mechanical parts—the fan for the power supply and the moving parts of the disk drives. You're unlikely to catch yourself in one of these. Peripherals are another matter, however. Printers in particular are loaded with moving parts. Most have a tag inside warning you to remove jewelry and to be careful of long hair. (Having it yanked out by an angry printer is no way to get a haircut.)

There is no such thing as being *too* safe. Just when you think that you've taken every possible precaution, look for something you may have forgotten.

PERSONAL SAFETY RULES

1. Probe carefully
2. Don't touch conductive surfaces
3. Observe the one hand rule
4. Use only insulated tools
5. Beware of jewelry, hair, neckties, loose clothing
6. Moving parts can be dangerous!

Williams: Repair & Maintenance of Your IBM PC (Chilton)

BEST RESULTS/MINIMAL TIME

COMPUTER SAFETY

Once you've taken the necessary precautions to protect yourself, you can begin to think about the well-being of the computer. As with personal safety, computer safety is basically a matter of common sense. Rules of thumb such as "Don't punt the computer across the room no matter how angry you get" and "Don't resort to your hunting rifle just because you're losing at Pac-Man" are obvious (or should be, although you'd be surprised at what some people have done to their computers). Other "don'ts" are just as obvious if you take a moment to calm yourself.

In certain ways the computer is a surprisingly tough piece of machinery. If it operates for the first week or so, it's unlikely that anything major will go wrong for many, many years—unless *you* cause it. Even if you do make some mistakes in operating or repairing the computer, chances are good that you won't do too much damage.

This doesn't mean that you are free to be careless. Just as you assume that all circuits are holding deadly charges just waiting to get at you, assume that any mistake will cause the immediate destruction of a $1000 circuit. (It *is* possible.)

Again, there is no such thing as being *too* careful.

PHYSICAL DAMAGE

The most common damage caused by the home technician is physical. Physical damage is also the least necessary. There is no reason or excuse for it. By being in a hurry, by losing patience, or from carelessness, the wrong thing is done at the wrong time and something snaps.

Most of the parts of your computer are tough. But others can be damaged all too easily. Caution is the key at all times, no matter how tough you think something is (or should be).

When removing a board, do so with slow and steady pressure. If it doesn't come out fairly easily, check to see if you've forgotten one of the holding screws before applying more force. The boards are supposed to be somewhat tight to maintain a reliable contact. But they're not in so tight as to require removal with both arms and a foot. If it doesn't move, there is usually a logical reason.

One computer operator decided that the boards were too difficult to remove. He took the probe of his voltmeter and jammed it into the slots of the receiving board to widen them. Then he was amazed that the operation of the computer was sporadic, at best. He ended up having to replace those receptacles at a premium cost.

Williams: Repair & Maintenance of Your IBM PC (Chilton)

FIG. 1–2   To remove an expansion board, grasp it firmly on both ends and lift. Don't forget to remove the holding screws! If it doesn't come out easily, there is a reason.

Another operator couldn't seem to get a board to come out. His fingers didn't seem strong enough, so he took out the pliers! After he broke off two chunks of board he discovered that he had forgotten to remove the screw that attached the board to the frame.

Boards aren't the only items held in place by screws. The drives are mounted securely in place. The designers realized that the drives are prone to jostling. Once the fastenings are removed, they should slide out easily. If they don't, you've missed something. (The same applies to reinstallation. The drive should slide easily into place. If it doesn't, the fault is your own.)

Never force anything. Take the few extra seconds to find out why the board, drive, or whatever won't move easily.

The components inside your computer have anywhere between 2 and 64 connectors. Each of these leads is prone to physical damage, mostly from bending them too far. The IC's are particularly sensitive. The ICs (or chips) have a number of metal prongs coming from them. More often than not the prongs on a new IC are in the wrong position for easy installation. Bending them manually is the usual solution for the computer owner who is installing chips. This

Williams: Repair & Maintenance of Your IBM PC (Chilton)

presents the danger of bending the prongs too far (or too little), and then having to bend and rebend the prongs.

Due to the thinness of the prongs, this bending and rebending creates stress on the metal. The prong may break off. The metal may also crystallize in such a way that the current flow changes, making a seemingly good chip useless.

USE THE PROPER TOOLS

The solution to this and to related problems is to use the proper tools for the job at hand. Two special tools are available to handle the delicate chips. One is made to put the prongs into the correct places for installation (an "IC installer"). The other is designed to aid in the removal of chips (an "IC extractor"). Both tools are fairly expensive for someone who plans to install or remove just one chip per decade. Both are inexpensive when you consider the cost per ruined chip.

These tools offer an additional safeguard. Some of the chips in the IBM are

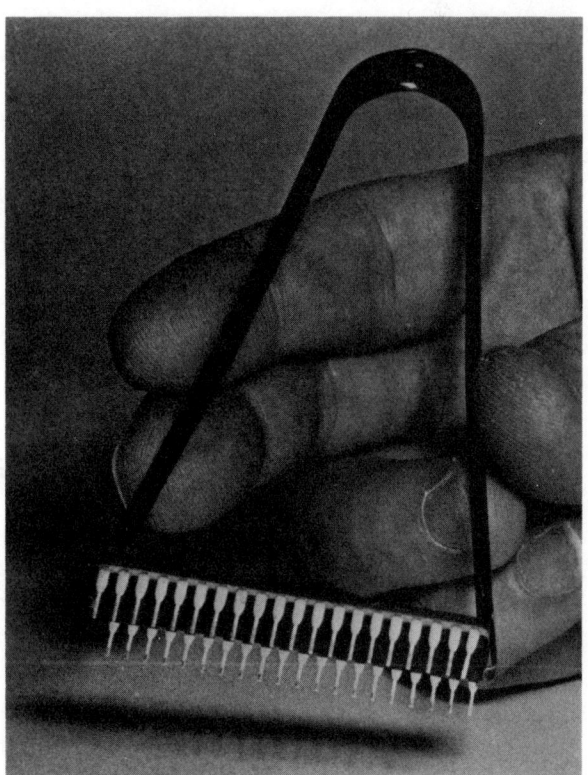

FIG. 1–3 The prongs of an IC are delicate and the inside can be ruined by static. *Handle with care.*

extremely sensitive to static. Your body has a tendency to store up static charges. You have probably experienced a tiny shock when touching a doorknob or other metal object after walking across a rug. If the charge was enough for you to feel it, it was probably enough to fry the insides of a delicate chip. Even a static charge you cannot feel at all may still be enough to ruin a chip.

The IC extracting and installing tools will help prevent this from happening. You can reduce static build-up by treating the carpets, either with a commercial product or with a dilute mixture of water and standard fabric softener. You can also make use of a "static discharge device." This is a small device connected to a ground (such as a neutral screw on a wall outlet). You touch a metallic spot on the device with your finger, and any static charge in your body is drained off safely.

SHORT CIRCUITS AND OHM'S LAW

The second most common type of damage caused during repair is a short circuit. This can happen in several different ways. The usual way is by touching the metal tip of a tool or probe across two points that are not meant to be connected. Most often this won't matter. Other times it will send a circuit off into oblivion with a cloud of smoke.

There is a mathematical relationship between voltage, amperage, and resistance. This is stated by Ohm's Law. The formula for this law is $E = I \times R$. The E is the voltage, the I is the current in amps, and the R is resistance in ohms. If you multiply the current times the resistance you'll know the voltage.

With some basic algebra you find that $R = E/I$ and $I = E/R$. By sticking some numbers into that last formula you can see what happens with a short circuit. It simply states that the current flowing is equal to the voltage divided by the resistance.

The voltage remains constant due to the design of the power supply. In most circuits this will be 5 volts. If the resistance is 10,000 ohms, the current flowing is .0005 amps. A short circuit effectively drops this resistance to near zero, which means that the current will flow to the limit allowed by the power supply (about 3 amps). For computer circuits designed to handle just fractions of a milliamp, the effect can be disastrous—like trying to instantly force a few hundred gallons of water through a tiny tube meant to carry just a few ounces.

In the previous section there was mention of necklaces and other jewelry causing current to flow into your body. While it's true that *inside* the computer there isn't enough current flow to cause harm to you, nevertheless the rule about jewelry applies—this time to protect the computer. If that necklace swings down and creates an electronic bridge, you may not be able to feel the effect, but the computer probably will. Resistance drops to near zero. As a result, current swings in the other direction.

The human body normally has a very high resistance. Unless your hands are wet, touching an active circuit inside is unlikely to cause any damage. (This means a digital circuit, not one of those that carry high voltage or ac.) A ring on your finger or a watch on your wrist is another matter. The metal will act the same as if you had connected a wire from spot to spot.

All these possible causes of short circuits are fairly easy to keep track of. Pay attention to what you're doing and you should have no problem. But other things can cause shorts that are more difficult to notice. For example, the cabinet is held together by screws. Each board is held in place by screws (through the protective metal covers for the expansion board slots). Both disk drives are secured by screws. Newer PCs have more screws yet, including some that are easy to miss. Worse yet, some boards and interface cables are held by a bolt and nut (instead of by a bolt alone). When you remove the bolt, the nut may fall off into the computer unnoticed. As you put the computer back together, the bolt will seem to hold things together by itself, making it easy to forget that the nut is supposed to be attached. Nuts and other holders fall off all too easily in disassembly. When you turn on the power, that lost and forgotten chunk of threaded metal becomes a surprisingly efficient conductor.

A less likely, but still possible, cause of electrical problems are stray bits of metal. The average computer owner won't have to worry about these unless a screw has been forced (which could strip off a piece of the thread) or unless the lead of a component has broken off.

So the rule of thumb is to work carefully and to check carefully for misplaced screws, nuts, and bits of metal before putting the cabinet back in place or turning on the power.

COMPUTER SAFETY RULES

1. Shut off power
2. Take notes, make sketches
3. Don't be in a hurry
4. Never force anything

5. Use the proper tools
6. Avoid short circuits
7. Check for screws, etc.
8. Beware of static

PREVENTING ELECTRICAL DAMAGE

To prevent accidental short circuits, flip the switch to cut off the power before removing anything. There is only one reason for the power to be on while working inside the computer, and that is for testing, probing, and measuring (and those must be done carefully). For anything else, the power should be off and should remain off. Make shutting down the power your automatic response, and make applying power what you stop and think about.

Imagine yourself going inside the the computer to remove a disk drive for a simple repair, such as to replace a broken door. You forget the rule and leave the power on. First, you disconnect the plug to the drive and there is a soft "zzzt." To get at the holding screws you have to pull any boards that are in the computer ("zzzt, zzzt, zzzt"). The door is replaced and you go to install it again. You've been careful about reinstallation and know that everything is where it is supposed to be. But nothing works. Each of the boards, including the one on the disk drive, has ruined components. You scratch your head and say, "Now how could that have happened?"

This isn't *likely* to happen. What is important is that it *could* happen. If a board or component is removed while current is flowing through it, the current value often changes. As the value changes in one place, other changes will take place elsewhere. Small changes probably won't cause damage, although they can cause the circuit to "age." Larger sudden changes have effects similar to those of short circuits or static discharge, namely they destroy the circuit from the inside.

When working inside the computer (with the power off!) pay attention to what you are doing. Look carefully at any connector you are going to remove. Make notes and sketches so you have something for reference. (Keep these notes and sketches handy for future times when you want to get inside the computer.) Before you turn the power on again, look around inside. Are all the connectors back in their proper places? Have you left any screws, nuts, or bits of metal inside? Are the accessory boards pushed all the way into their slots?

COMPONENT REPLACEMENT

Usually a component failure is best handled by replacing an entire board. Even professional technicians use this method of repair. It may sound as though this should be more expensive than just replacing the component. Actually it is not. Tracking down a problem to a single component is time-consuming, and often the time involved ends up costing more than a new board.

Since you probably do not have the equipment that the professional technician has, nor the knowledge to use it, you will probably confine most repairs to board replacement rather than component replacement. (When you handle the repair by replacement, keep in mind that the malfunctioning board or unit has a trade-in value. A board that cost $300 new won't, or shouldn't, cost $300 to replace.)

There will be times when you will be able to identify the exact component and will want to replace just this piece. When this is possible, the savings to you are large. Most of the components in your computer cost very little. Diodes, resistors, and capacitors cost just pennies. The ICs are often just a few dollars. Many of the ICs in the PC can be purchased for under a dollar.

The first thing to do when replacing a component is to make certain that the new component is exactly the same as the one it replaces.

Some components have polarity (positive and negative ends). An electrolytic capacitor, such as the filter capacitor in the power supply, has two leads, one positive and one negative. Install a new one backwards and it could explode (yes, explode). Other components won't react so violently but could cause damage throughout a circuit. Expensive damage. ICs that are not installed correctly can burn up and may take a whole string of other components and circuits with them.

Even when you aren't replacing a component, pay attention to polarity. In many repairs you'll be disconnecting cables and other wires. Most of these have special keyed plugs, making it impossible to reconnect them with the wrong polarity. But you may run into a few that don't have this intelligent design.

Taking notes and making sketches are important parts of any repair. Get into the habit, even when you don't think that you'll need the notes and drawings. There is no need to be a professional artist or writer. What you do is primarily for your own use, a jog to your memory. Assume, though, that others will be using the notes. It's possible that you'll have to consult a pro on the repair. The notes you take, and the drawings you make, could save you quite a bit of time and money, regardless of how rough they are.

SOLDERING

Some components plug into place. Most are soldered into place. Despite what you might think, soldering is an art. It's not the kind of skill you can learn in a few minutes. With circuits as critical as those of a computer, you certainly shouldn't be practicing inside your PC. There are a number of books and pamphlets available on how to solder. Heath Company offers a course in soldering. Before you even consider soldering inside the computer, learn everything you can about it. Then practice, practice, practice, on something else first. Learn to solder correctly and with the proper type of soldering tool, or don't solder at all.

The soldering iron used for digital circuits, such as those in the PC, should be rated at no more than 30 to 40 watts. Anything hotter could too easily damage the circuit or the board. Even with the lower-powered iron, anything more than a few seconds of contact is risky. Many components are very heat sensitive. The internal goodies can be fried all too easily. The board may also be damaged permanently. If this happens, you may as well scrap the board and buy a new one.

The soldering iron should be designed specifically for digital soldering. These irons are more expensive, but the extra cost is a necessity. They have grounded tips, which prevent damage from any build-up of electrical charge. *Don't* try to use any old soldering iron or gun for the job.

PREPARING TO WORK

Before you begin any repair you should have a solid understanding of the correct procedure. Diagnosis and repair is a logical sequence of steps (more on this in Chapter 2). Learn these steps. Follow them! They'll save you money, time, frustration, and a whole lot of trouble.

The first step is so simple that most people ignore it. Make back-ups of every bit of software. Many books and courses suggest this and mention making *a* copy. Instead, make at least *two* copies. Having the original plus the two copies helps to protect you in a number of ways.

If you don't know how to make a copy, refer to "Copy" and "Diskcopy" in your *DOS* manual. The manual also tells you how to check the accuracy of the copy. Even after such an accuracy test, run each copy to be sure that it works properly or that it contains the correct data. *Always* test at the time of making the copy. If you don't, you won't know for certain if your copy *is* a copy or is just a useless diskette.

There are losses through fire, water, forgetfulness, and so forth. These should always be of concern to you, however unlikely they might seem at the moment. You must also consider loss through the machine. In a large computer repair house a technician fed in the diagnostics disk. There was a problem with the disk drive. The result was that the machine ate the diskette. His solution was to feed in a second diskette. The computer destroyed this one as well. So he booted up a third, then a fourth. By the time he figured out that the drive was annihilating the recorded programs, five copies were destroyed.

This may sound silly. A professional *should* know better. The average operator may not. If the problem is intermittent, it would be very easy for even the most experienced operator to waste a program or two before realizing what the problem is.

If your machine doesn't accept the original, check it with a copy. If this copy doesn't work, chances are good that something is wrong. If the problem has erased the first two, you have the third to protect yourself.

WHERE IS THE PROBLEM?

Most computer malfunctions are operator or software errors.

Computers are extraordinarily reliable. Most people have come to think of machines as being at fault when something goes wrong. In many cases this is true. If a car suddenly stalls on the freeway, it is usually a machine error. If a television or radio refuses to work, the fault is usually with the equipment. With a computer, the fault is often with the person running it.

Most appliances are designed so that *anyone* can operate them. There aren't many things for the operator to change or vary. A television, for example, has very few controls. The owner can switch it on, change the channel, and change the contrast and color within limits. Beyond this, about all the television operator does is to sit back and view.

Operating a computer normally involves pushing a variety of buttons, each of which does a different task. It's like having thousands of controls available instead of just a half dozen. But the increased number of controls offers many more opportunities for making mistakes.

Before you tear into your computer, make sure that the fault lies with the computer. Chances are good that the fault is your own or is the fault of the programmer. Most malfunctions are not the fault of the computer.

One of the first questions you should ask yourself is "Has it *ever* worked?" An untried program may have flaws. Even a known program may have bugs in it. A tried and true program may give out after a number of uses. (Diskettes are well manufactured, but aren't without error.) Then there are those programs

Williams: Repair & Maintenance of Your IBM PC (Chilton)

WHERE IS THE FAULT?	IMPORTANT QUESTIONS TO ASK YOURSELF
1. Operator 2. Software 3. Peripherals 4. Disk drives 5. Computer	1. Has it ever worked? 2. Has that function of the program ever worked? 3. Do other programs work? 4. What is working, and what is not?

that you use every day for their normal functions, but that have other capabilities you haven't yet used. When you get around to trying out the other capabilities, you're back to a "Has it every worked?" situation.

When a problem occurs, the first thing to suspect is yourself (or the operator of the computer). Software documentation is notorious for being poorly written. Do you (or does the operator) understand how to work the program? Are you trying a new function of a familiar program? Have you read through the instruction manual completely?

Second in the line of suspicion is the software itself. If you have back-up copies, try one of these. (If you have been using the program successfully and have tested the copies for all functions, you can eliminate the "Has it ever worked?" question.) Another test is to try a different program. For example, if your word processing program isn't working, try your accounting program or one of the games you have. Use the DOS diskette if you happen to have only one program.

If you have checked out any possible operator and software failure, the next step is the visual check. Look for what doesn't look right. Don't start pulling boards and components until you've completed this step. Look for the obvious. Much of the time you can solve the problem with very little effort. Perhaps you've missed a screw or nut and this is shorting one of the internal circuits. Maybe the door to the disk drive is cracked and won't allow the spindle to make proper contact with the diskette. (An unfortunately common problem.)

The next step is to make use of the diagnostics diskette. Each time you turn on the machine it checks itself. (The people at IBM were really thinking ahead on this!) If the system doesn't pass this automatic mini-test, go to the diagnostics diskette provided with the computer. (A more advanced version is available through IBM. If you want it, check with your local dealer or write directly to IBM for the current price.)

Williams: Repair & Maintenance of Your IBM PC (Chilton)

FIG. 1–4 If something doesn't work, the operator is usually at fault. Check the *Guide to Operations* manual, then try another diskette, then use the diagnostics diskette.

Notes should be taken throughout your testing. Even before you stick in the diagnostics diskette you should have something written down. What *is* working? What *is not?* If the "Print Screen" command is still working, let your printer run off a copy of the error codes, etc. that the computer kicks out.

Don't waste your time on parts that are functioning properly. Diagnosis is a process of elimination. If you know for a fact that the drives are accepting programs, that your printer is operating from "PrtSc," that the monitor is giving a correct image of what is being sent to it, then eliminate or ignore these sections of the computer. If the problem is that the RAM won't hold data, why waste time taking apart the drive? All you will do is cause more problems.

Look for the most obvious and work your way down.

All checks begin with the cabinet closed. Make a note of what is happening or is not happening. Check and recheck for operator error, then for software error or for diskette failure. Only then should you think about opening the cabinet. You should have at least a fair idea of what you're looking for before opening the cabinet.

When this step comes along, move slowly and deliberately. With the box

Williams: Repair & Maintenance of Your IBM PC (Chilton)

open look around for the obvious. Don't just start tearing into the computer. You will cause damage if you're in too much of a hurry.

PREVENTING PROBLEMS

Chapter 7 deals with maintenance. Your IBM has been designed and built to require very little maintenance. In fact, you can pretty much ignore your machine, almost to the point of abuse, and it will keep going.

There are still some TLC requirements (you know—tender loving care).

The greatest enemy of the computer is dust. A tiny fleck of dust that your eye can't even see can gouge a diskette to uselessness. Dust combined with humidity can cause short circuits. Yet dust is everywhere. All you can really do is to reduce the amount that gets into your computer, particularly into the mechanical parts. Some dust on the boards is unlikely to cause any problems. But just a few invisible particles on the disk drive heads can slice the data on a diskette to shreds.

One item you can eliminate entirely is food and drink. Make a policy never to allow anything spillable within a 20-foot radius of the computer. If you or some other operator wants a cup of coffee, it's time for a break—*away* from the computer. Liquids in particular are dangerous. Spilled into the keyboard they can necessitate total replacement of the keyboard, plus possible repairs inside the computer caused by the short circuits.

SUMMARY

When in doubt—*don't*. That's the shortest way to put it.

Computer *do's* and *don'ts* are little more than common sense put into practice. If something seems silly or risky, don't do it. If it seems logical and sensible, think it over before you do it.

A computer is a logical machine. Things don't go wrong for no reason.

ENEMIES

1. Dust   5. Other Contaminants
2. Liquid   6. Humidity
3. Food   7. Weight or pressure
4. Smoke   8. Carelessness (i.e., YOU)

Williams: Repair & Maintenance of Your IBM PC (Chilton)

BEST RESULTS/MINIMAL TIME

There is always some reason. Just because you can't see it right away doesn't mean that the simple, logical cause isn't there.

Don't attempt to do anything unless you have some idea as to what you are doing and how to do it. Likewise, *don't* attempt a repair without the proper equipment.

In making repairs, keep in mind that the design of the machine demands exact components. If a resistor goes out, the replacement for that resistor must have exactly the same value.

Suspect yourself first. Next, suspect the software and diskettes. After this, begin all checks with the cabinet still closed. Take notes constantly. Make sketches where applicable. Don't trust your memory, because you will find it unreliable.

Finally, *do* read through the entire applicable chapter before you attempt to work on a section of your computer. If you suspect the drives, for example, read Chapter 3 thoroughly before you begin. (Read Chapter 2 even more thoroughly before you do anything!)

Repair and maintenance of a computer is not all that big a deal. People less intelligent than you are doing it every day and without making errors. At the same time, people with more intelligence are messing things up faster than they can be repaired—almost always because they refuse to follow "the rules."

===

Diagnosis: What's Wrong with It?
2

When something goes wrong with your computer, it's tempting to immediately remove the cover and start poking around. Even if you have some idea as to what has happened (or has not happened), this is the worst way to begin. The cure of a problem *always* begins with the cabinet closed, and usually with the power off.

Diagnosis is a step-by-step process. Often you can skip certain steps. When you do, you should know that you're skipping them and why. (This will be because the diagnostic step has taken care of itself automatically. A step such as checking to see if the computer is plugged in obviously does not have to be performed if power is getting to the computer.)

The primary diagnostic steps are covered in this chapter. Once you have tracked a problem to a particular system or device by using these steps, you will be guided to the correct section of the book for further diagnosis and for the final repair or replacement.

For example, imagine that your computer refuses to accept a program. The cause could be many things. This chapter will take you through a diagnosis until you have tracked the problem to a single part of the computer system. If the problem is in the software, you will be instructed at that point to turn to Chapter 3 for more details. If the preliminary diagnosis indicates that the problem is caused by the disk drives, you'll be directed to Chapter 4.

It's as simple as that. When a problem comes up, begin right here in this chapter (unless you already know for sure what is causing the problem). Diagnosis is little more than a process of isolating the cause of the problem. By using this chapter you can eliminate many of the things that *are not* causing the problem. You can then more easily pin down what *is* malfunctioning.

BEFORE OPENING THE CABINET

Most problems and malfunctions can be taken care of without ever taking the computer apart. Many can be spotted and cured without even turning on the power.

There are six steps before you open the cabinet:
1. Check for operator error.
2. Check for software error.
3. Look for the obvious.
4. Take notes and make sketches.
5. Do the power-on self test.
6. Use the diagnostics diskette.

CHECK FOR OPERATOR ERROR

A computer has remarkably few hardware malfunctions. Despite its appearance, the computer is relatively simple—much less complex than, for example, your television set. If something goes wrong with the television set, the chances are good that the fault lies in the set. After all, there is little chance of operator error. Even the new programmable television sets require little on the part of the operator compared to handling a computer.

The computer works because of what the operator does. It has hundreds of controls, generally accessed through the keyboard. The more controls you are required to operate, the more likely you are to mess up somewhere along the way. Then add to this the accidental flubs, such as pressing the wrong key, and you begin to appreciate the differences between handling a computer and the operation of a television set.

If you are the operator, much of the time you'll know when you make a mistake. If the operator is someone other than yourself, this may not be true. It's possible that the problem could get worse as the operator tries to recover from the error, making your job of tracking it down more difficult.

Your own first BASIC programs are excellent examples of how important operator error can be. Each command has to be just right. A program gives directions to the computer and guides it through the complex electronic maze inside. Give it incorrect directions and the computer will get "lost" and will indicate an error.

Don't think that operator error can only occur with "home brew" programs. Even software that has been professionally written and produced isn't free of suspicion. (See the next section.) A flawless program can present some very strange troubles if you don't understand its functions, characteristics, and quirks.

Does the operator know how to operate the program? Is it a new program, or perhaps a new feature of a familiar program? (Remember to ask, "Has the program or function *ever* worked?") If either the program or feature is new, then it's possible that the "malfunction" is nothing more than a lack of knowledge on the part of the operator.

If you could spend some time in a computer repair shop and listen to some of the malfunctions (and their solutions), you'd come to realize just how many things the operator can do wrong. It has nothing to do with being stupid or even careless. Most of the time the problem is due to an honest mistake. One operator was never shown how to start the machine, let alone how to run it after it was going. Another had the power cord kicked out by the family dog and couldn't figure out why the computer seemed dead. Still another erased a large amount of valuable data because he thought that the diskettes had to go through "Format" each time before loading them.

If there's even the remotest possibility that the problem is operator error, check it out completely before blaming the computer. Of all computer "malfunctions," about a third are brought about by nothing more than operator error. (See Chapter 9.)

CHECK FOR SOFTWARE ERROR

Once you've eliminated the operator as the source of the error, be sure that the software isn't the cause. This includes both the data on the diskettes and the diskettes themselves. Both can produce errors that may seem to be machine problems. Of all problems that come in to repair shops, the vast majority are brought on either by operator error or by software error. In a sense, the software becomes a sort of operator once it's fed into the computer. It tells the computer what to do and how to do it when the human operator pushes the various keys.

New programs and diskettes are especially suspect. Just because the box and plastic wrapper are intact doesn't mean that nothing could have happened to it. In some ways the diskettes are as delicate as Christmas tree bulbs (see Chapter 3). Despite all the care and testing, a flaw could have snuck in during manufacturing, or the diskette could have been damaged in transit.

A program on the diskette might have been imperfect to begin with, or may have faulty sections. (I have a chess game for my PC in which the king cheats whenever he's in trouble.) Newer programs are more open to suspicion than programs that have been around for a long time. After several thousand users run the program through its paces and find the errors, the manufacturer can release an improved version. (The fact that new programs often contain weaknesses of various sorts isn't necessarily the fault of the manufacturer, although it is likely.)

LOOK FOR THE OBVIOUS

A new PC owner took his system home, pushed in the program diskette (just as he'd been shown at the shop), but nothing happened. That same afternoon he tucked all the equipment in his car and brought it back in. It operated flawlessly, so he took it back home again—only to have the system refuse to operate again. Next day he was back in the shop.

"I just don't understand it," he said. "I know the outlet is good because I plugged a lamp into it. Maybe something got jiggled inside the computer when it was in the car." Again the system operated perfectly while in the shop, and with the owner standing there watching. Then he saw the technician reach back to flip off the power. "What's that red paddle for?" he asked.

That may sound silly, but it is a true story. Somehow he managed to get the idea that shoving in the disk automatically kicked in the power. It's such an obvious thing that the salesman hadn't even bothered to show the customer how to apply power to the computer.

This same customer might have been tempted to take off the cabinet to see what was the matter. It probably would have done no harm. On the other hand, he could have caused actual damage before he realized that all he had to do was flip a switch.

Look for the obvious before you do anything else. If the computer seems dead, look to see if the plug is still in the outlet and check to make sure that the power switch has been flipped before tearing apart the power supply.

The same applies to all cables and connectors. It's easy for them to become loose even if your computer sits perfectly still. You can't always tell if the connector is secure just by looking, either. Push them in to make sure that contact is being made.

Contrast and other controls on the monitor might have been bumped or accidentally turned so that something *seems* to be wrong. The more people there are who touch your system, the greater is the chance that something has been bumped, nudged, or otherwise messed up by human action.

A program refusing to load could be due to something as simple as your having accidentally inserted the disk upside down or having put in the wrong diskette. A flickering on the screen or recording error could be caused by someone in the next room turning on a vacuum cleaner or an electric mixer.

Then there are problems with the physical construction due to normal wear and tear. What might appear to be a major problem with a disk drive could be nothing more than a broken door. (They're plastic and all too breakable.)

Even inside the cabinet, keep your eyes open for the obvious. There are more connectors inside the cabinet. The boards may not have been pushed all the way into the expansion slots. A screw may have fallen to cause a short. Faulty components are sometimes obviously damaged: a capacitor might be leaking fluid, a resistor might be obviously burned, or a soldered connection might be loose.

As you're going through the more detailed steps of diagnosis, keep looking for the obvious. Start with the simple, obvious things and *then* go to the more complex.

CHECKING FOR POWER

If all connections seem sound and still nothing happens, it's time to do the first obvious step—check the incoming power. Checking for power in an outlet is easy. The easiest method is to plug something else into the socket, such as a lamp. If the lamp lights, you know that there is power coming in through the outlet. It won't tell you much more than this, though.

Using a meter is a more accurate gauge. Set the meter to read in the 120 volt ac range. The gauge will tell you not only if power is coming in, but it will tell you how *much* power there is. Power companies are famous for producing "dirty" power. It has periodic drops and surges. The problem is compounded at times of peak power demand. In the middle of a hot summer afternoon, for example, the power company may be having a hard time keeping up with the demand placed on the lines by the thousands of air conditioners. Line voltage is bound to drop.

The power supply in the PC will easily tolerate any voltage between 104 and 127 volts. If the voltage goes beyond these limits, a built-in safety circuit will shut everything down. Even at the computer's 104 volt minimum, your test lamp will probably work just fine. The outlet will *seem* to be just fine, but the computer will still refuse to function. The gauge will show you why.

Note: The fan in your PC is wired directly to the incoming 110-volt line. If it is operating, then the wall outlet and the power cable are good. The problem is in the power supply.

MAKE NOTES AND SKETCHES

Throughout the diagnosis, take notes, and lots of them. What is happening? What is not happening? What symptoms are you noticing? If you have to take the computer to a technician later on, these notes will save you time and money. Even if you fix it yourself, the notes will help to guide you along.

The notes should include sketches. This way you'll have an easier time reassembling things after you're finished. Don't trust your memory. It's too easy to forget that screw #17 fits into such-and-such slot to hold this particular thing to that particular thing.

The plugs and connectors are keyed. (This may not be true for certain devices from other manufacturers.) Even so, keep track of which connector goes where. The more you disassemble, the more important will be the notes and sketches.

POWER-ON SELF TEST

The people at IBM have really started something! When they designed the PC they built in an automatic self test. Each time you switch on the power, the computer gives itself a quick check through to make sure that everything is working properly. Other computer manufacturers are now incorporating the same idea. (Wouldn't it be great if your car would go through an automatic diagnosis each time you started it up?)

For some strange reason, there are companies around that have developed special adapters to defeat this self test. The idea behind it seems to be, "Why should the operator have to wait 30 seconds? Let's just by-pass that self test so they can get going faster."

Apply power to your computer a thousand times and you are unlikely to find a single problem from the mini-test. Bypassing the mini-test doesn't seem to be such a bad idea. Not until you think about it.

The power-on self test (IBM calls it "POST") runs quick checks on the system board (sometimes called the mother board), the RAM memory, the monitor, the keyboard, the diskette drives, the fixed drive (if you have one), and the expansion chassis (if you have one). Note, however, that it does not test the drive heads, and that the keyboard check is only for stuck keys.

With minimum configuration the self test takes just 13 seconds. If you've filled your computer will accessories and additional RAM, it will take longer, since there is more to check. With a fully loaded system the check will take about 90 seconds.

The more options you have, the more important it is to let the self test run itself. If your computer has 512K of RAM and two 320K drives, you can't really take the chance that nothing is wrong in there. Imagine spending three or four hours punching in data, only to find that the RAM has gone bad, or that the disk drives won't write. All of a sudden that 90 seconds seems like a very short investment in time.

NORMAL RESPONSE

Normal response is for the cursor to blink on the display, followed by one short beep, followed by the "IBM Personal Computer" display. If you're loading in DOS (or any diskette with the system command on it), the display should show this. If no diskette is in the drive (or if the drive is malfunctioning—see Chapter 4), it will give the "Personal Computer Basic" display. (If the program being loaded has an *autoexec.bat*, it will perform the self test and go immediately into the program. You will not get the usual IBM DOS or BASIC display.)

ERROR RESPONSE

If something has gone wrong, the self test will give audio responses and will display codes on the screen to guide you to the malfunctioning part. (See Table 2–1.) For example, a continuous beep indicates that the power supply is faulty (Chapter 6). One long beep followed by a short beep tells you that the system board (Chapter 5) is causing a problem.

The codes are arranged in a logical sequence. For example, any code in the 100s means that the fault is in the system board. Anything in the 200s indicates memory. The 300s are for the keyboard, and so on.

If the error is so severe that it cannot be displayed on the monitor, you will rely on the audio code. Make a note of these and refer to Table 2–1.

DIAGNOSTICS DISKETTE

Tucked in the back of the *Guide to Operations* manual is a diagnostics diskette. This is designed to help you spot most problems and to run periodic checks. By using this diskette, and the information contained in this chapter, you'll be able to track down most of the common failures within your system—often down to the individual components.

Run the diagnostics diskette when the machine is new and in perfect operating condition and again after any additions. This will show you what the response is *supposed* to be.

Knowing how to use this diskette correctly can save you many frustrating hours. Why waste hours trying to determine the problem when you have the

TABLE 2-1
Power-on Self Test Error Responses

Audio Indication	Problem	Refer to Chapter
No beep (no display)	Power	6
Continuous beep	Power	6
Repeating short beeps	Power	6
1 long, 1 short	System board	5
1 long, 2 short	Monitor	6
1 short, no display	Monitor	6
1 short, BASIC on screen	Drive	4

Code Indication	Problem	Refer to Chapter
101, 131	System board	5
201	Memory	5
xxxx201 & Parity Check x	Memory	5
Parity Check x	Power	6
301, xx301	Keyboard	6
601	Drive	4
1701	Fixed drive	6
1801	Expansion unit)	5,6

tool to do so right there in front of you—your own computer? Most machines lack the ability to tell you what's wrong. Your computer *has* this ability—*if* you know how to use it.

The diagnostics is somewhat like the self test, and again uses codes. (See Table 2-2.) The major difference is that the diagnostics diskette is considerably more powerful and will perform the tests in more detail. It will also record the errors if you wish. However, be aware that the diagnostics does not check some external devices, such as modems. It may also give an error reading if it is testing a board or device manufactured by someone other than IBM, or if the external device is connected to the adapter card.

When the diagnostics diskette has been booted (loaded), the screen should appear as in Figure 2-1, with four operations from which to choose. Options 1 (Format) and 2 (Copy) are functional and also provide an easy way to check certain operations of the disk drives and the keyboard without going to the DOS diskette. (These programs on the DOS diskette are better, however, and should be the ones you use for normal operation. Full explanation of their use is in the DOS manual.) By selecting option 9 you will exit the diagnostics routines. For diagnostics checking you would select option 0.

Note: If you are going to test the "Formatting" program on the diagnostics

TABLE 2–2
Diagnostics Error Codes

Code	Problem	Example	Refer to Chapter
02x	Power	021	6
1xx	System board	100	5
20x	Memory	203	5
xxxx	Memory	1471	5
xx20x	Memory	29203	5
30x	Keyboard	300	6
xx30x	Keyboard	34309	6
4xx	Monitor (b&w)	403	6
5xx	Monitor (color)	505	6
6xx	Drives	601	4
7xx	8087 coprocessor	702	5
9xx	Printer adapter	900	5
11xx	Asynch comm.	1101	5
12xx	Asynch comm. (alt.)	1200	5
13xx	Game adapt	1303	5,6
14xx	Printer	1405	6
15xx	SDLC comm. adapter	1510	5
17xx	Fixed disk drive	1701	4,5
18xx	Expansion unit	1801	4,5,6
20xx	BSC adapter	2021	5
21xx	BSC adapter (alt.)	2121	5

Note: If the last two digits of the code are zero, the system checked is operating correctly.

diskette, be *sure* to remove that diskette and insert a blank one *before* trying to format.

After entering a 0 (for diagnostics), the screen should appear similar to Figure 2–2. There will be differences in the display, depending on the equipment you have. For example, if your computer has 64K in RAM, the number by "KB Memory" will be different. If you don't have an asynch communications adapter, it won't be listed on the screen.

This first step verifies that the computer knows which devices you have connected. One of the ways it does this is by checking the two DIP switch settings inside the computer. If the display on this screen is different from what you have connected, it's time to check all connections and cables again and the settings on all the switches. (See "Helpful Tables and Charts" at the end of this book for switch settings.)

Setting the two DIP switches merely tells the computer what is *supposed* to

```
The IBM Personal Computer DIAGNOSTICS
Version 2.03 (C)Copyright IBM Corp 1981, 1983

SELECT AN OPTION

0 - RUN DIAGNOSTIC ROUTINES
1 - FORMAT DISKETTE
2 - COPY DISKETTE
3 - PREPARE SYSTEM FOR RELOCATION
9 - EXIT TO SYSTEM DISKETTE

ENTER THE ACTION DESIRED
? 0_
```

FIG. 2-1 Screen display when the diagnostics diskette is first loaded.

```
        THE INSTALLED DEVICES ARE

        S   SYSTEM BOARD
        S   192KB MEMORY
        S   KEYBOARD
        S   MONOCHROME & PRINTER ADAPTER
        S   2 DISKETTE DRIVE(S) AND ADAPTER

        IS THE LIST CORRECT (Y/N) ? y_
```

FIG. 2-2 Screen display after selecting option 0 (run diagnostic routines).

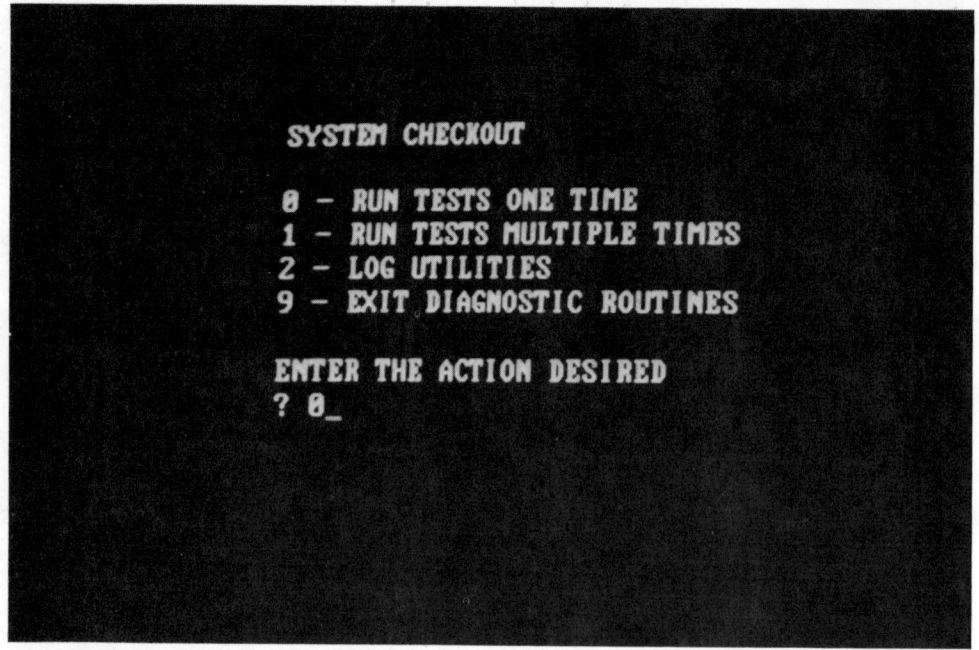

FIG. 2–3 Running the diagnostic tests one time (option 0) is the normal choice.

be there. If they are set incorrectly, you'll get some confusing errors. For example, if you have 64K in RAM, but the switches tell the computer that you have 256K, the test of the memory will indicate that there is a failure in that section.

If everything matches enter "Y" to indicate that things are as they should be. This will bring up the next display (Figure 2–3). Again you are given four options: 0 (to run the tests once), 1 (to run the tests more than once), 2 (record the utilities), and 9 (exit).

OPTIONS 0 AND 1

Option 0 will be your normal choice. It takes you step-by-step through the tests and checks every system in the computer one at a time. The procedure is given in the *Guide to Operations* manual. Follow this and you should be able to pinpoint the trouble area.

The diagnostics routines will test the system devices and options one by one, beginning with the system board (100) and through all the options up to and including the BSC adapter (2100) if you are one of the rare individuals who has one. (See Figure 2–4.) As a unit is tested, the display will either tell you that the unit is operating correctly (by showing the two zeros at the end of the code) or that it is malfunctioning (by a code other than one with the two zeros at the end). If diagnostics shows that a particular unit is faulty, make a note of the error code and continue through the diagnostics. Other errors might occur in other units. Only after you've made notes on *all* the errors that the computer finds should you proceed to the other chapters in this book.

Option 0 (test one time) requires a response from you on most steps. The keyboard test, for example, asks you to press each key and watch the display for the correct symbols (Figures 2-5 and 2-6). If the unit being tested passes, you are told to enter "Y." If the screen doesn't match and you enter an "N," an error code will be displayed. Make a note of this before going on.

Option 1 (multiple testings) does *not* require a response from you during the diagnostic steps. That is, it removes your ability to tell the computer that something is wrong on one of the displays. You're relying on the computer to catch all errors. This option has an advantage in that it stands a better chance of catching an intermittent problem. As you begin this option you are given the choice of how many times to run the tests (1 to forever) and the option to have the computer stop at each error. This allows you to take note of the error codes and make other notes as well, if you wish.

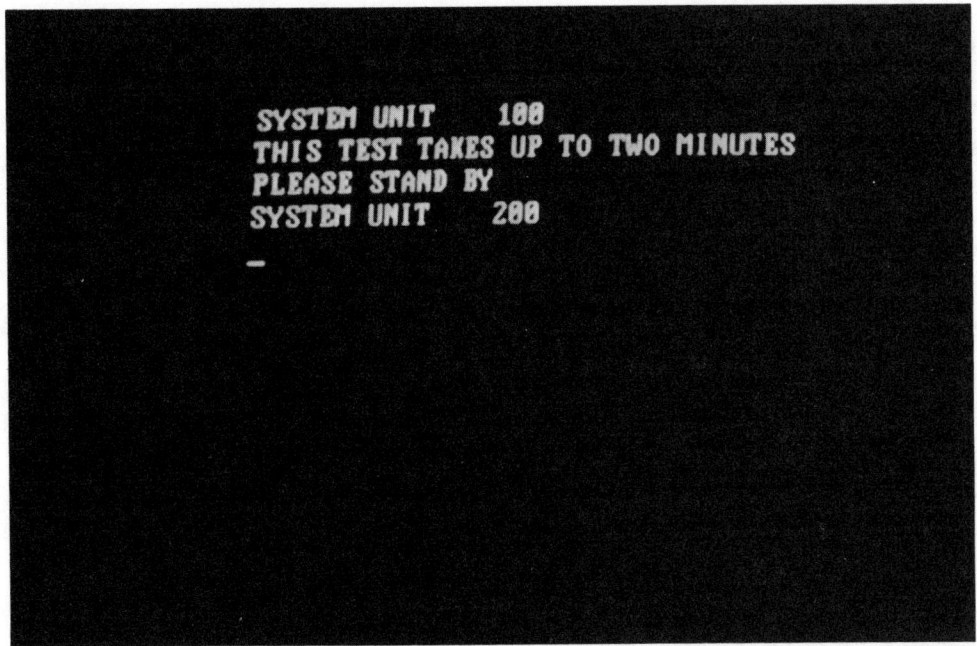

FIG. 2-4 The diagnostic routine tests all the system units one by one.

Williams: Repair & Maintenance of Your IBM PC (Chilton)

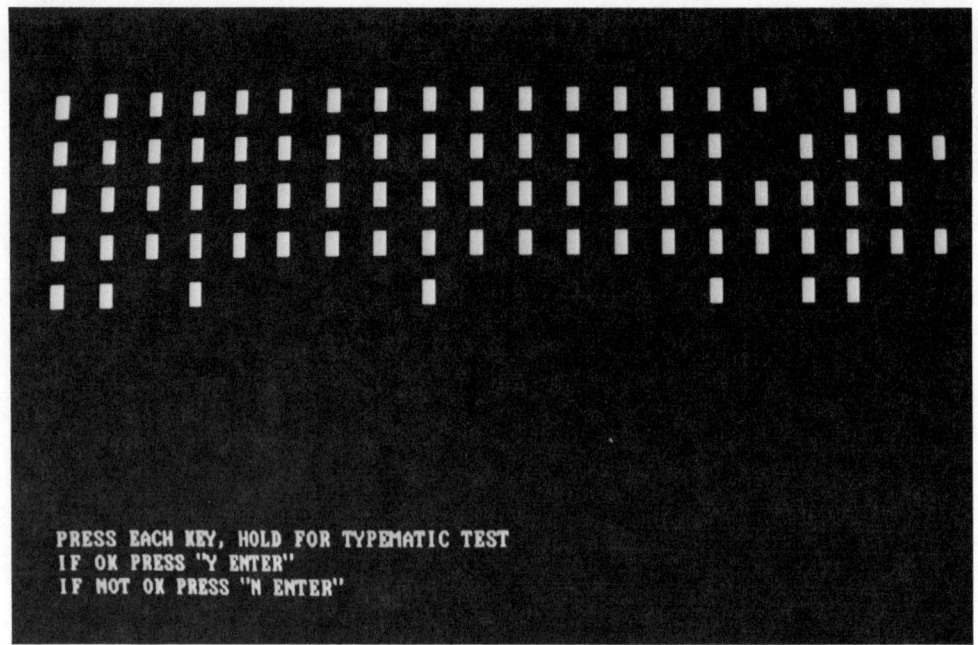

FIG. 2-5 Keyboard test display prior to testing.

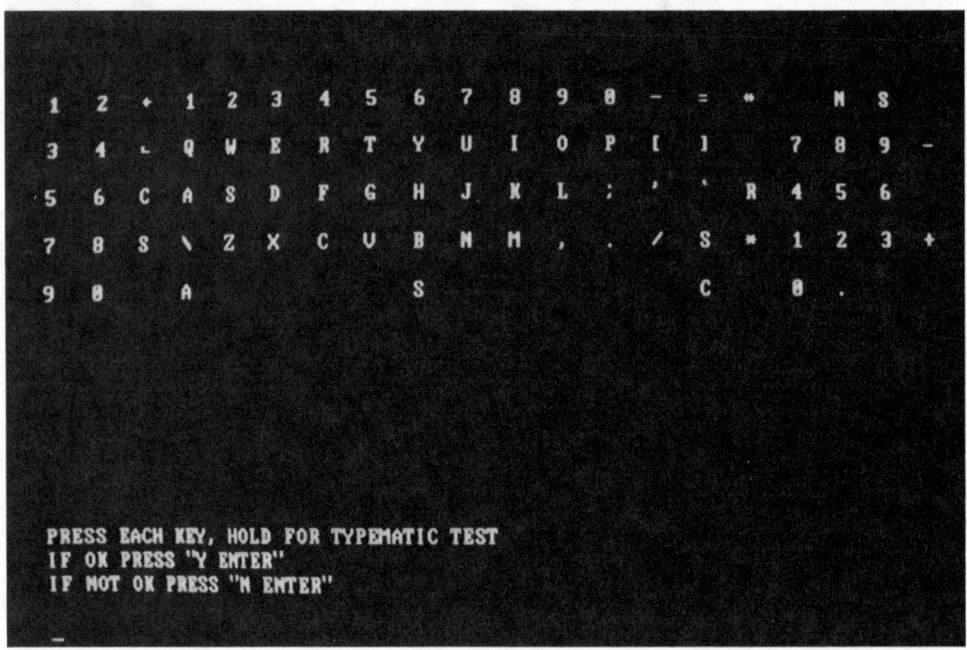

FIG. 2-6 Keyboard test display after testing.

OPTION 2—LOGGING UTILITIES

Both standard and advanced diagnostics give you the opportunity to record error messages as they occur. This can be done through the printer, to the diagnostics diskette, or to the cassette (which is strange because I've never met anyone who has a cassette hooked up to a PC). This is made possible by the "Log Utilities" option.

Logging the errors to diskette requires that you have a copy of the diagnostics in drive A that is not write-protected. Operating the program in this manner requires that you stay with the computer. When the program reaches the part where it tests the drives, scratch (blank) diskettes must be inserted (see Figures 2–7 and 2–8). This means that you'll have to remove the diagnostics diskette each time the drive test comes around and replace it with a blank diskette.

Logging the errors to a printer can be run without your supervision. You can place a blank diskette in each drive since neither is used except for the drive testing. You can then run the "Multiple Testing" option.

AFTER DIAGNOSIS

Once you've completed all the steps from the diagnostics diskette, you should have on hand at least one error code (assuming of course that something is mal-

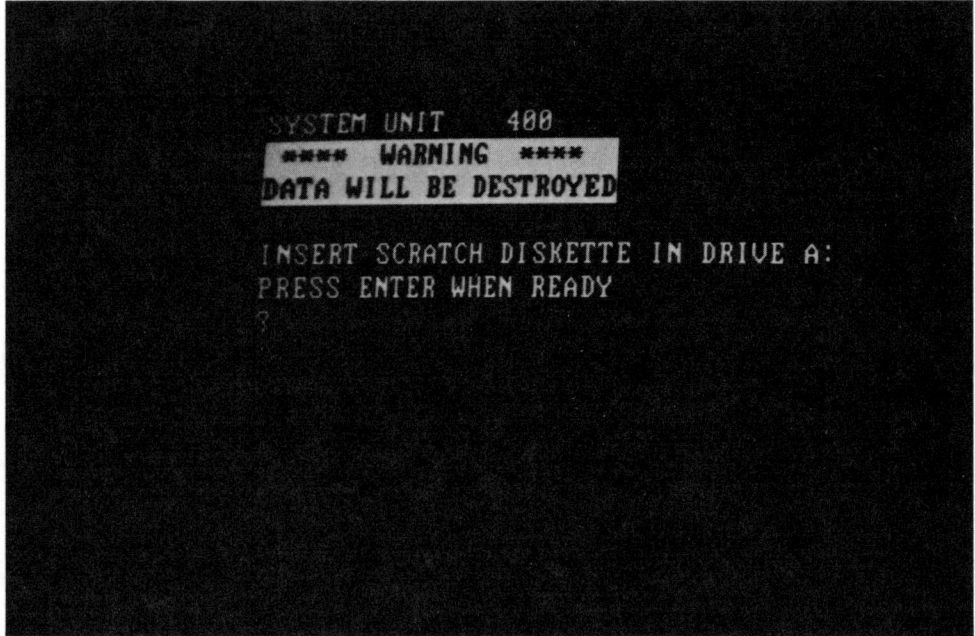

FIG. 2–7   Screen display prior to drive test.

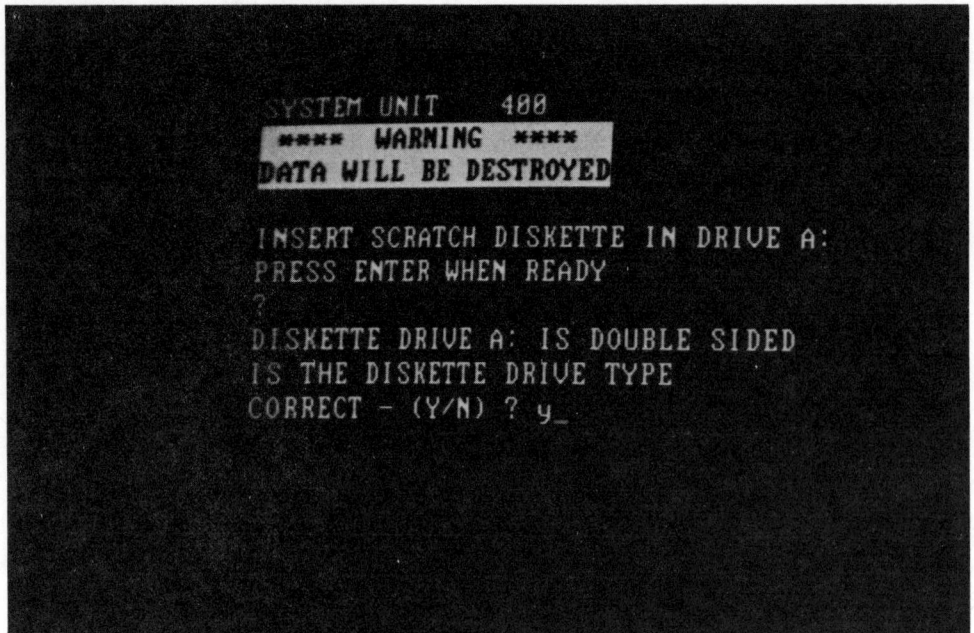

SYSTEM UNIT   400
**** WARNING ****
DATA WILL BE DESTROYED

INSERT SCRATCH DISKETTE IN DRIVE A:
PRESS ENTER WHEN READY

?

DISKETTE DRIVE A: IS DOUBLE SIDED
IS THE DISKETTE DRIVE TYPE
CORRECT - (Y/N) ? y_

FIG. 2–8 Screen display after drive test.

functioning). Refer to Table 2–2. This will guide you to the proper chapter for further diagnosis and testing.

SUMMARY

Diagnosis is a matter of listing the possible causes, and then eliminating those things that are not at fault until you find the one or two things that *are* causing problems. This is not nearly as difficult as it sounds. You have all the "tools" you'll need already.

Most problems have nothing to do with the computer or its devices. By careful observation, you should be able to find out if the malfunction is within the computer, is the fault of the operator (most common cause of trouble) or is within the software (second most common cause of trouble).

For mechanical or electronic problems, the IBM PC was designed to diagnose most problems itself. The power-on self test (POST) will automatically run preliminary tests on the various systems and devices each time you apply power. If the POST tests don't supply sufficient information, the diagnostics diskette that came with your computer almost certainly will. There are very few malfunctions that this won't spot for you.

The PC will tell you exactly what is wrong, and where. There are error codes you can follow. Some are displayed only with an advanced diagnostics package (available from your dealer or IBM at an additional cost), but most appear either during the POST or the standard diagnostics tests.

While your system is operating correctly, go through all the testing procedures provided in the diagnostics diskette so you will know what things are *supposed* to look like.

Anytime something seems to be malfunctioning, take notes. Make drawings if you do any disassembly. Both will guide you along and will provide valuable information if you have to consult a technician.

Once again, the steps in diagnosing a malfunction are:

1. Check for operator error
2. Check for software error
3. Visually check for the obvious
4. Use power-on self test
5. Run the diagnostics diskette
6. Be sure to take notes and make drawings throughout
7. Repair or replace when you can
8. Consult a technician when you can't, armed with all the above information to save time and money.

====================================

Diskettes and Software
3

If you were to tell a repair technician that your computer was malfunctioning (assuming that the problem wasn't obviously something like a power supply failure), he would immediately try to find out two things. First, what was the operator doing at the time? Second, is the software functioning? (For that matter, has it *ever* functioned?) Many "malfunctions" are due to operator or software error.

Operator error occurs for many reasons. Even the most experienced operator can make a mistake now and then. The more complex the program is, the more likely it is that the fault is with the operator. Before you complain to the software or hardware manufacturer, eliminate all possibility of operator error. (Don't be too surprised if you find that the fault is yours even when you're *sure* that it is not.)

Ask yourself, "What have I done wrong?" Then ask, "Has it *ever* worked?" Don't answer either too quickly. After all, would you rather spend hours and hours tinkering with a machine, or a few seconds to be honest with yourself?

Go through the manual and other documentation again. These materials are notorious for being poorly written. (I know several people who have erased $500 programs due to confusing instructions in the "Installation" sections of the manuals.)

New programs are always suspect, both for software error and for operator error. If you've never tried to use the program before, you may not be putting in the proper commands. (Back to that lousy manual again.) Or you might be using a particular feature of the program for the first time. (Back to the manual.)

Williams: Repair & Maintenance of Your IBM PC (Chilton)

All clear? The fault is *definitely* not yours but in the program? Fine. Now we can proceed.

DISKETTES

The usual method of data input and data storage with the PC is with 5¼″ diskettes. They are often called "floppies" because of their flexible nature, and are sometimes called simply "media." If you think about what a diskette is and what it does, it might seem strange that they don't cause even more problems.

Information can really be packed onto the surface of the diskette. Each byte is made up of eight bits (or pulses), yet each byte takes up little more than one ten-thousandth of a square inch. (The space used to store a byte is about one hundredth of an inch wide and a few thousandths of an inch in length.) Just as your IBM won't accept a DOS command with a character missing, it probably won't accept a program with a scratch or blockage, even if the damage is less than a thousandth of an inch in any direction if it is over a critical spot.

HOW DISKETTES ARE MADE

The diskette begins as a thin sheet of flexible plastic (Mylar is the standard). (Mylar is a Dupont company trademark. The generic name is polyethylene terephthalate.) The plastic comes to the disk manufacturer in rolls that are about a foot wide and often about a mile long. The rolls are tested, inspected, and cleaned.

Next the plastic is given a magnetic coating on both sides (even if the diskette is later given the "single sided" label). This coating is made up of extremely fine magnetic particles, a binder (like glue), and a lubricant. The microscopic particles have to be "glued" to make a perfectly uniform coating across the surface of the plastic. If they are not, there will be gaps and data will not be accurately recorded or read.

Once all this is done, the coated plastic is smoothed and placed back on rolls where it is given a number so the manufacturer can keep track of it. It is then stamped into the disk shape and is polished (called burnishing) again. The rougher the surface is, the more damage it will do to the diskette read/write heads. A diskette with a rough surface can also cause the read/write head to skip, just as a particle of dust can cause the needle of a record turntable to skip. The result is lost data and potential damage to the heads. It isn't possible to eliminate all sources of wear, but efforts to reduce wear are made by all the better manufacturers.

Williams: Repair & Maintenance of Your IBM PC (Chilton)

The finished diskettes are placed inside the square outer covering (usually made of PVC plastic). Inside this jacket is a layer of thin and very soft material which helps to keep the surface of the diskette clean (Figure 3-1). Without this layer the diskette would be constantly "attacking" the read/write head with particles of dust and other contaminants. The liner helps to protect the read/write heads by gently cleaning dust away. It also protects the diskette surface by preventing the soft diskette from rubbing against the harder plastic jacket.

Tests are run throughout the manufacture of the diskette. Its final label (single sided, double sided, single density, double density, quad density, etc.) is determined by these testings. A diskette that passes all the tests is given the highest rating, and carries the highest cost to you.

What this means in simple terms is that the least expensive diskettes (single sided, single density) have the same basic surface and manufacture as the most expensive diskettes. They've just failed some highly sophisticated test along the way and the manufacturer doesn't want to guarantee that the diskette will accurately hold data in higher densities. Even so, you will probably be able to format them as if the diskette were of higher quality. It's a fairly common practice for computer owners to try to save money by buying the less expensive single-sided diskettes and use them as double sided.

The problem with this is that you don't really know why the diskette was given the "single sided" label. Even if the diskette formats correctly for double

FIG. 3–1 An unjacketed diskette. The small hole is the index hole.

Williams: Repair & Maintenance of Your IBM PC (Chilton)

DISKETTES AND SOFTWARE

sided use, and "Chkdsk" (see your DOS manual if you are not familiar with this term) seems to show that the diskette is good, you might go in one day to retrieve some valuable data only to find that it has disappeared due to a disk flaw.

Diskettes are the least expensive part of your system, and are also one of the most important parts. If you try to save money by buying cheap or poor-quality diskettes, you're almost sure to lose in a larger sense.

ANATOMY OF A DISKETTE

Diskettes used for the PC are called "soft sector." This means that the computer's formatting program divides the diskette into the correct number of sectors. The sectors begin in a certain spot on the floppy which is determined by the index hole (the small hole located near the hub access hole). Since sector size is determined by the formatting program of the particular computer brand and model, the same diskette can be formatted for use in different computers.

A "hard sector" disk or diskette has a series of index holes to mark the sectors. These are pre-set by the manufacturer and such diskettes are good only for certain machines. (Actually a soft sector diskette does have a number of index "holes" to mark the beginning of the sectors. Most of these are invisible magnetic spots on the diskette.)

For the IBM PC there are 40 tracks (numbered 0 through 39). These are like the grooves of a record except that they are concentric circles and not in a spiral. When you hear your disk drive grind it is because the read/write head is moving from track to track. (The sound you hear is the head stepper motor causing the read/write head to move. For more information on this, see Chapter 4.) Standard track width is about one hundredth of an inch, with normal density being capable of holding about 6000 bits per linear inch.

Each of these tracks is divided into eight sectors (nine sectors for DOS 2.0), with each sector capable of holding 512 bytes (a byte is a character or other piece of data that contains eight pulses) of data. Each track, then, can hold 4096 bytes of information. One side of a diskette can hold 163,840 bytes (40 tracks × 8 sectors × 512 bytes). To make things easier, this is called 160K. A drive that uses both sides of the diskette holds twice this, or 327,680 bytes (320K).

Note: Formatting under DOS 2.0 gives the diskettes nine sectors instead of eight. A diskette formatted under 2.0 will have nine sectors of 512 bytes each on the 40 tracks—or 9 × 512 × 40 = 184,320 bytes per side. A double-sided diskette would then have 368,640 bytes of storage available. For ease, this is usually called 360K.

In the center of the diskette is a round hole about an inch in diameter. This allows the spindle in the drive to make contact with the diskette and spin it.

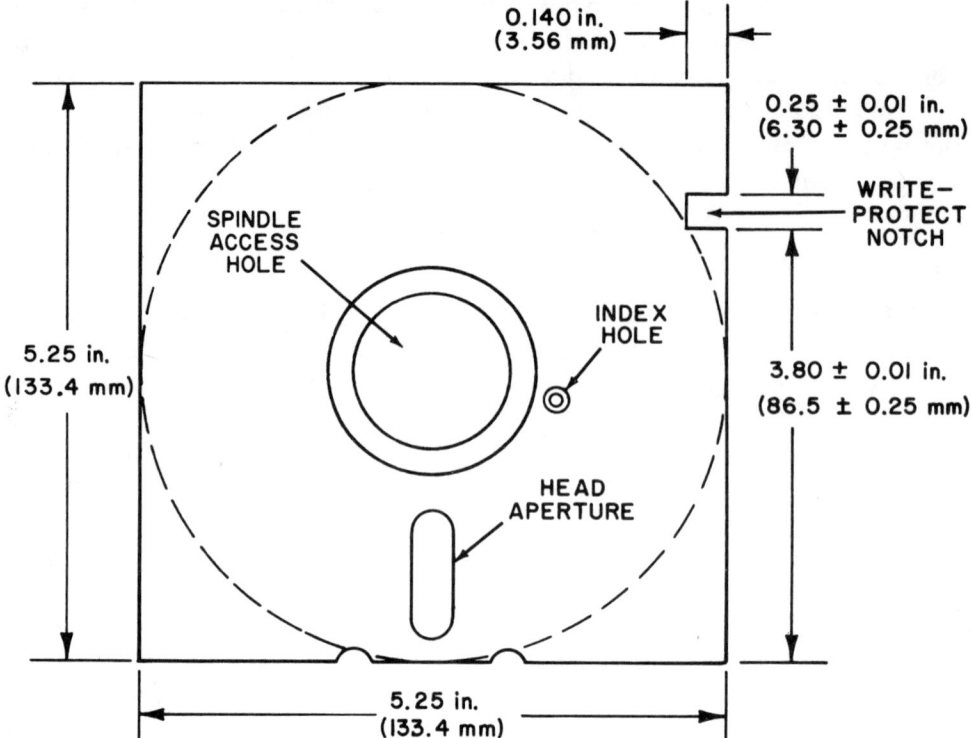

FIG. 3-2 Anatomy of a diskette.

Most diskettes also have a band of extra material called a hub ring. This ring both protects the diskette from damage by adding strength where it is needed most and allows a better contact between diskette and spindle.

On the jacket, very close to the spindle access hole, is the small index access hole. Somewhere on the diskette itself is a small puncture. As the diskette spins, the hole passes by an index sensor and tells the drive where the first sector is on the track. (You can rotate the diskette by hand and see this hole.)

The oblong cut from the diskette cover allows the read/write head of the drive to get at the information. This is the most sensitive part of the entire diskette. A fingerprint here can cause all sorts of troubles, both to the data and to the read/write head in the drive. (According to one major manufacturer, at least 80% of all diskette failure is attributable to fingerprints.)

When putting new labels on the diskettes, it is very important to keep both the index hole(s) and the read/write access holes open and uncovered. If either is covered, two things will happen. First, the diskette won't work. The machine

Williams: Repair & Maintenance of Your IBM PC (Chilton)

will give a "Not ready" error. To the computer there is nothing in the drive. Second, the glue from the label may come off onto the recording surface of the diskette. If this happens, you may as well trash the diskette. You cannot clean a diskette, nor should you try.

Along the side of most diskettes is a small notch. This allows a small switch inside the drive (see Chapter 4) to activate the recording head. When this notch is covered with tape, the recording head cannot function, and you cannot write information onto the diskette. (You can read from the diskette, however.) It's a good habit to cover this notch on any diskette that has data you won't be changing.

You can get diskettes that have a write-protect notch on both sides of the diskette (or you can cut them yourself). Such diskettes will also have a second index hole. When these extra notches are present, the diskette is usually called a "flippy floppy," which means that you can use the diskette as though it was two single-sided floppies. To make use of the second side all you have to do is to turn the diskette over. The second write-protect notch and second index hole allow the drive to look at the second side of the diskette as a new one.

These flippy floppies are used more often with single-sided drives. If you have single-sided drives in your PC, a flippy floppy allows you to make use of all 360K of storage available on the diskette. On a double-sided drive, the second side is already available. Even so, a flippy floppy has an advantage here. You can have a program of up to 160K on each side, complete with an *auto-exec.bat* to have the program boot up by itself. This won't work with a normal double-sided diskette. You cannot repeat a file name on a diskette. (This has to do with the way the formatting program allocates the files.) The second *auto-exec.bat* will cover over the first, the same way an edited file will replace its original.

A flippy has the disadvantage of wearing out sooner. As it is used on one side it rotates in one direction. When you flip it over, it rotates in the opposite direction. This causes wear and may increase the possibility of scratches on the surface from particles captured by the lining.

(You can buy devices that cut both the write-protect notch and the index hole to change a standard diskette into a flippy floppy. Unfortunately, the PVC jacket tends to shatter when cut, which can leave particles of the PVC on the diskette.)

HOW DELICATE IS A DISKETTE?

Despite its apparent complexity, the diskette is surprisingly tough. Many professional technicians have stories of playing catch with an unjacketed diskette, and then have it perform flawlessly. Diskettes that have been almost shredded by deep scratches zip through the drive as though brand new.

At the same time, a tiny piece of dust could cause a diskette to "crash" and become useless. You're never sure which will happen. Nor is lost data the only risk. Each speck of dust can be ruining the read/write head while it is slowly grinding the magnetic coating from the diskette surface.

The disk drive of the PC reads and writes 48 tracks per inch (48 TPI), which means that the 40 tracks used for data, plus the empty spaces between the tracks, are squeezed into just $\frac{5}{6}$ of an inch (which includes the gaps between the tracks). On a normal double density drive, the 163,840 bytes per side are packed into about 6 squares inches of surface area. A quad density drive (96 TPI) doubles the amount of information crammed onto the diskette.

Now you can see why the diskettes and drives are so sensitive. The read/write heads record or retrieve information from tracks that are about one hundredth of an inch wide and separated from each other by about one hundredth of an inch—all this while the diskette is spinning merrily at about 300 rpm. If the accuracy is off by just a slight amount, or if something gets in the way, the data on the diskette may be inaccessible.

CARE OF DISKETTES

You can't easily "repair" software. If your business program is malfunctioning, you won't be able to get inside to fix it. (Some programs can be fixed, such as those you've written yourself or which are written in a language you can use, such as BASIC.) Your goal is to prevent problems before they exist.

Software problems can be greatly reduced by simply taking care of the diskettes. The disk is tough but not indestructible. It's also unpredictable at times. One day you can play catch with the diskette and have it work. The next day a speck of dust or cigarette smoke could fall onto the diskette and wipe out everything. Not only can you lose software and stored data, you can cause damage to the disk drives by not properly caring for the diskettes.

Care of diskettes is not complicated, nor is it time-consuming. The manufacturers have taken great pains to ensure that the diskettes will last for a very long time, and with a minimum of problems. Extensive testing is done before the diskettes are sold, and the diskettes are made to be as tough as possible. It's not uncommon to find a manufacturer who guarantees that the diskette will not fail even after several million passes per track. In time this translates to nearly a year of constant running before the life expectancy is reached. Since normal operation calls for the diskette to be running just seconds out of every operating hour, the disk should, and could, last a lifetime.

Williams: Repair & Maintenance of Your IBM PC (Chilton)

DISKETTES AND SOFTWARE

```
DISKETTE SPECIFICATION STANDARDS

Tracks per inch                 48 TPI
Tracks (number)                 40
Track width                     .0108-.0128 inch
Track density                   6000 bits per inch
Temperature (operation)         50 to 112°F
                                10 to 44°C
Temperature (storage)           −40 to 140°F
Humidity (operation)            −40 to 60°C
Humidity (storage)              20% to 80%
Disk speed                       5% to 95%
                                300 rpm
```

(The lifespan of a diskette varies, however, with how it is used. Although the manufacturer might guarantee the diskette for three million passes per track, each time you use the diskette it goes to the "Directory" track. Some applications refer to this same track many times. Thus, the life of the diskette depends largely on how many times this one track is used.)

The diskette can withstand any temperature between 50° and 120° F (10° to 50° C) and still operate without error. Even if the temperature happens to go beyond this range, the diskette is still likely to recover if you give it enough time to cool down or warm up. (See "Heat and Cold" on page 54.)

Humidity does little actual damage. The official humidity range for a diskette is between 20% and 80% (5% to 95% for storage). Drier environments tend to dry out the diskette (although it *does* take quite a while) and, worse, cause static build-up, which can alter data. Higher humidity can cause dust to stick to the diskette and the liner to swell. If this happens, the diskette may not spin properly and you'll get an error display.

PHYSICAL DAMAGE

The soft liner inside the jacket does more than just keep the diskette clean. It also serves as a cushion for the diskette to prevent damage. It does a fine job for most normal things, but it can't protect the disk against everything. This is up to you.

Anything that puts pressure against the diskette can cause a dent in the jacket, in the liner, or in both. At best the diskette will have a hard time spinning in the drive. If this is all that happens, you may have enough time to make a copy of the ruined floppy.

This is why a felt tip pen is suggested for writing on the labels. Better yet, write on the label before sticking it onto the diskette. The tip of a ballpoint pen or pencil could easily damage the diskette. Even a soft felt tip can press a groove into the diskette. Although the pressure you use in writing may not seem like much, all that pressure is concentrated at the tip of the pen or pencil and is being applied to the diskette surface through that tip. The force is effectively multiplied because of this.

Having weight of any kind pressing against the floppy can cause damage. Diskettes are best stored vertically (standing up). This not only protects the diskettes, but it reduces the risk that you might drop a stack of books on top of a diskette.

The problem of weight or pressure is further compounded if dust or other particles are trapped in the liner. Imagine having the pressure of five pounds of books concentrated onto the sharp edges of a piece of dust a millionth of an inch wide. Dust may *seem* to be soft. To get an understanding of what it can do just take a look at the windows of a car left in a dust storm.

Pretend that the soft liner of the diskette is made from coarse sandpaper. You *know* you wouldn't risk the information stored on such a diskette by placing even a small amount of weight on it.

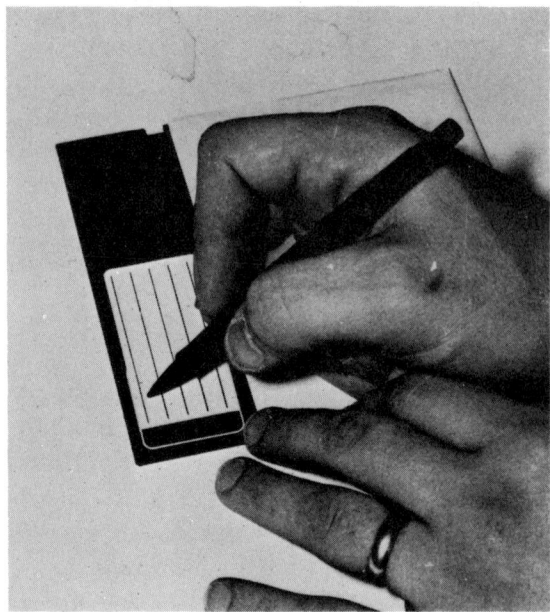

FIG. 3–3 If you *must* write on the diskette, use a soft felt-tip pen.

Williams: Repair & Maintenance of Your IBM PC (Chilton)

STORAGE

When not in use, store all diskettes in their cover jackets and preferably standing vertically inside a box. This is to reduce the accumulation of dust and other particles. The soft inner lining will help to protect the read/write heads, but it tends to capture particles which in turn can cause scratches on the diskette. With the data being so tightly packed (163,840 bytes per side), even a small scratch can have devastating effects. That scratch may occur on an unimportant part of the diskette. It might also happen over a critical bit of data and make the rest of the diskette useless.

A quality storage box might cost $30 or more. This sounds expensive until you think of what you're protecting. Most PC owners have hundreds of dollars invested in software, and perhaps thousands of hours spent in punching in data. It's not unusual for a computer owner to have more invested in software than in the computer system itself. Why take the chance of throwing all that down the drain just to save a few dollars?

If you can't afford to buy a diskette storage box, make one. Such a box should *not* be made of metal (because of magnetism). Use wood or plastic or even cardboard. (Plastic is best because cardboard and wooden boxes can have large amounts of dust in them, no matter how well cleaned.) The inside should be clean and unpainted (fumes). The top should close tightly enough to seal out dust. Beyond that it can be as fancy or as simple as you wish. (I know people who use modified shoe boxes with great success.)

However you do it, keep dust and other particles and contaminants to a minimum. You've invested too much time and money to waste it on a lousy environment.

If a diskette gets dirty, *do not* attempt to clean it. Your cleaning is virtually guaranteed to cause more damage than any amount of dust. How dirty the diskette is and what kind of contamination it has will determine what you do with it. If it isn't too bad, store it for severe emergencies. Otherwise toss it out. A dirty diskette means that it's time to pull out one of the back-ups. (Make another back-up before going to work.)

MAGNETISM

The data on the diskettes is stored magnetically. It should be obvious that the diskettes must be kept away from other sources of magnetism. Yet computer operators are constantly erasing their valuable programs and data by forgetting the obvious. Some cases of forgetfulness are as blatant as setting the diskette next to the magnet of a speaker. Most involve more subtle sources.

Inside the telephone is a small electromagnet. Normally it just sits there and does nothing. But whenever someone calls, that little device lets fly with enough magnetism to destroy a diskette. (It rings the bell in the telephone.)

Other potentially dangerous sources of magnetism are the monitor, the printer, the modem, the cabinet of the computer, any tape recording machine, fluorescent lights, and even a calculator. Motors work by using magnetic fields. If you're not sure, don't trust it. (Anything metal is automatically suspect.)

HEAT AND COLD

More important than solar radiation is the heat generated by the sun. Since the diskette is usually black, it tends to absorb more than its share of heat. Leaving it in the open sunlight is very likely to cause damage. Even if it's close to a normal incandescent lamp, it could pick up enough heat to cause damage.

Keep the diskettes away from all sources of heat, sun and other. At best heat can warp the jacket of the diskette. If this happens, the data you've recorded won't be in the same place on the diskette. It won't matter, though. If the diskette or jacket become warped, the diskette probably won't spin in the drive anyway.

The same things can happen with cold. Not only can extreme cold cause the diskette to crack, it can cause the recorded data to shift in position. A sudden change in temperature from cold to hot can cause other problems as well. Even a slight change in temperature, and the contraction or expansion this causes, can cause the tracks to move away from where they are supposed to be. Keep in mind that the tracks are just barely over a hundredth of an inch wide, and that each byte of data covers a mere ten-thousandth of a square inch.

BACK-UP COPIES

Although it is not strictly maintenance, ALWAYS make back-up copies of important programs (if copyable) and data. The cost of diskettes is low considering their value to you. Making back-ups is the least expensive method there is to protect yourself against software failure. Make at least two back-up copies of all software and data diskettes that are important to you. The more important the original is, the more back-ups you'll want to make. If the data on those back-ups is no longer important, you can always reformat the disks and use them again. In the meantime you'll be protected.

Both the "Chkdsk" and formatting programs are good guides as to possible problems if certain sectors are bad.

While diskette manufacturers guarantee the physical quality of their diskettes, these guarantees *do not* cover the data on the diskettes. If a diskette goes bad, the manufacturer will replace it with a new diskette of the same kind, but the data is lost forever, as is the time you spent in punching it in. Back-up copies are a kind of insurance you don't want to be without.

PROGRAM PROBLEMS

If you're writing your own programs, you're almost bound to make some mistakes. The IBM will generally tell you that you have made a mistake, and it will even show you the line(s) where the mistake was made.

If your program doesn't work, accuse yourself before you accuse the computer. Refer to the DOS and BASIC manuals to make sure that the commands you've punched in are correct. If you like programming, take some courses in the subject. Get some books. Learn the most efficient (and the correct) ways to do things.

When it comes to purchased software, you have much less control over what has been done. Many programs are inaccessible for corrections. Unfortunately, so are all too many software companies. It isn't uncommon for a company to have a disclaimer in the package which says in effect, "If the program doesn't function as promised—tough! You bought it, now it's your problem." Other companies support their products, but at an additional cost. (The old "If you can't understand our poorly written manual, pay us an extra $100 and we'll explain it to you" attitude.)

At other times you might be pleasantly surprised at the response. There are companies that do everything possible to make sure that the end user is happy and satisfied with the program.

When contacting a company about a software malfunction, be fair to them. Begin by doing everything you can to make sure that the error isn't your own. Read the manual carefully and thoroughly. If the problem still hasn't been solved, be as specific as possible in your communication to the company. Give them as many details as possible, specifying what you've done to correct the problem and the particular version of that piece of software that you own. They can't give much of a response to "It doesn't work—how come?"

Accurate notes on what has gone wrong and what you've done will be of help to both you and to the company representative. Notes can also get you in touch with otherwise inaccessible persons, since they make you seem to be more of a concerned expert. The more information you provide, the quicker will come the solution to the problem. Even jot down the page numbers in the instruction manual so you can readily refer to the proper section(s) for that particular function. *Don't* call a manufacturer until you have all the facts straight and immediately at hand.

One of the advantages of working with a local dealer is that you have a quicker access to information. Even if you haven't purchased a training course on a particular piece of software, they'll probably be happy to answer questions for you. Most will replace defective software without any hassle. (Bring your sales receipt! You can't expect them to guarantee a program that was sold to you by someone else.)

FAILURE TO BOOT

There are a number of reasons why a program will fail to load. The most common reason is a bad diskette. It may also be the fault of the drive, the power supply, the memory, or even the keyboard. Usually it is quite easy to find out what is causing the problem.

Eliminate all the obvious things first. Is there any power at all? (If the plug is in the wall and the outlet has been checked and there is still no power, go to "The Power Supply" in Chapter 6.) Are you using a new program and operating it improperly? Or perhaps the diskette you're using is very old and has simply worn out. Try another diskette, one that you know is good. If this one loads, you'll know that the fault lies with the software. (Use a diskette that is not critical. Although it is rare, it is possible for the drive to malfunction and write over the top of a diskette, even if it is write-protected.)

It's possible that the diskette you're trying to load is too large for the memory you have in the computer. Usually this will be displayed. Sometimes it will not.

I was trying to get a particular program to load, without much luck. It loaded to a certain extent but would stop and give the A> prompt. The fault was obviously in the software since other programs loaded normally. The computer being used had 64K. The program was slightly less than this and should have loaded without any problem. There was no obvious sign that the memory was too small. The problem turned out to be with the size of the DOS installed on the software. DOS 2.0 was used, which is somewhat larger than the earlier versions. The difference was just enough to cause the software to fail in loading. With DOS 1.1, the program loaded and executed perfectly.

If some disks boot and others do not, run the diagnostics diskette. (See Chapter 2.) This should show you if the fault is in the software, in the memory, in the drives, or somewhere else. For example, diagnostics might indicate that the drive is single sided, while you know it is double sided (or vice versa).

Before tearing anything apart, look for the obvious. Is the door to the drive intact? (The usual response from the computer if this is the problem will be to display the "IBM BASIC.") Inside, are the cables firmly attached to the drives?

A simple test of the drives involves switching the cables. Remove the cable from each drive and install them again in reverse order (i.e., the connector that was on drive A gets connected to drive B; the connector for drive B goes to drive A). For more information on the drives, read Chapter 4.

OTHER PROBLEMS

There will be times when a program loads and operates normally, only to malfunction while the program is running. Data may suddenly come out changed or missing. The display might show a "Parity Check" error. The program could lock up the keyboard, causing the loss of the data you've been punching in.

If the problem is in the software (in the program itself), you should be able to reproduce the malfunction by pressing the same keys again. You may have already noticed where the failure occurs. (Don't forget to check the manual and eliminate the possibility of operator error!) Notes will come in very handy in tracking down the problem.

Changed or garbled data can often be the result of overediting. The computer will automatically assign a chunk of data to a spot on the diskette. If possible, it will record these chunks in sequence. If something else has been placed in the next open spot on the diskette, the data will be moved along until a spot is found. This tends to break the file up all over the diskette. In reading such a broken file, the computer might miss something.

The solution for this is to occasionally perform a "Copy" on those files that are frequently edited. This command will rearrange the file in sequential order and help prevent a file from being scattered across a diskette. You can make a copy of a single file, or you can use "Copy A: *.* B:" to copy the entire diskette. (Do not use "Diskcopy" for this. "Diskcopy" makes an exact duplicate of the original diskette, including broken files.) It is best to use a freshly formatted diskette for the copy, since a diskette with data on it could break up the files even more (to make it fit between the existing files).

One track on the diskette is set aside for "File Allocation." Each time you bring up the directory of the diskette, the computer goes to this file and displays the files. Each time you tell the computer to load in a program or data file, it again goes to the allocation table to find out where the needed file is.

Earlier in this chapter we talked about how many passes a diskette can withstand before malfunction (three million passes per track). This seems as though the diskette could indeed last forever. It *will* last for many years. The main reason it wears out is due to the passes against the allocation track. Each time the file is read, recorded, or used in any way, the drive head goes back to the allocation track. After a few years of this, the track might fail. Despite the fact that the information is still good on the rest of the diskette, it is difficult to get at it because the allocation track no longer tells the computer where to look.

This same track can become faulty for other reasons. Some programs do not "exit gracefully." If the program is in use and you lose power (purposely by shutting down, or accidentally in a power outage), the allocation table can become messed up. The end result is about the same as if the track had worn out.

The solution for both is prevention. Make back-up copies of everything important. The more you use a particular diskette, the more important back-ups are. If a diskette has been in use for a long time, make a copy and replace it *before* it gives out.

SUMMARY

Most computer malfunctions are caused by either the operator or by the software. Eliminating operator error is a matter of proper training and of paying attention. The simplest beginning is to read the instructions. Learn how to work with the program and how to handle its functions—and its quirks.

Since you are unlikely to be the author of your functional programs, you cannot eliminate software problems. If the package comes to you with flaws in it, there won't be much that you can do, other than to return the package for a refund. As soon as possible after getting a new program, test it out. The longer you wait to do this, the more difficult it will be to get a refund or an exchange.

Make back-ups of all important software and data diskettes. Two back-up copies of each is the minimum. It's an inexpensive insurance against loss through operator goof-up, diskette flaw, or drive malfunction.

Taking care of the diskettes is simple, since there is nothing to do other than to prevent damage. Provide a clean environment. Keep the diskettes away from things that could damage them, such as magnetism, heat, contaminants, and physical dangers.

Handle them properly and they can last a lifetime. When they finally wear out, you always have the back-up copies to turn to.

===

The Disk Drives
4

The electronics of a computer allows electrons to move in a particular manner. The only motion is that of the electrons, and virtually the only wear is that caused by the heat developed. Something mechanical is bound to have more troubles than something that doesn't move physically at all.

The only mechanical parts of the PC are the fan, the printer, and the disk drives. Of all failures in your computer system, most will involve either the printer or the drives. (Printers are covered in Chapter 6.)

The disk drives are critical parts of the computer system. Without them there isn't much you can do with your PC. Although the PC is set up to take a cassette, the cable needed to plug a cassette machine into the computer is not easily available from any manufacturer. (If the disk drive malfunctions in certain ways or if the drive door is left open, the PC will automatically load "Cassette BASIC."

Note: Both the keyboard and cassette ports in the back look the same. Be sure that the keyboard is plugged into the correct one or the computer will not operate correctly. If your computer does not have the labels in place, the cassette port is the one furthest from the power supply.

The only way to load in programs other than "Cassette BASIC" or to save *any* program is to have a disk drive. (You *can* keyboard in programs in BASIC under "Cassette BASIC" and they will operate. You just can't save them.) Without the drive your PC is little more than a very expensive toy.

DRIVE TYPES

There are two basic drive types used in the PC. Many of the steps for diagnostics and repair are the same for both. A few things are different. Before you go through the following steps, you should know which drive type you have. You can tell by the name printed on the circuit board or by the serial number.

Probably the most common type is that made by Tandon (Figure 4-1). It has a serial number that begins with A, B, or with no letter prefix at all, and it is called "Type 1." A "Type 2" drive (Figure 4-2) is made by Control Data Corporation (often called a CDC drive). Its serial number begins with the letter D. Drives from other manufacturers are available. If you have a drive that is not Tandon or CDC, some of the diagnostic steps described below will not apply.

One of the major differences between the Tandon and CDC drives is in the way the lines are connected to the drive. The Tandon (Type 1) uses a number of separate plugs labeled 8 through 13. The CDC has a single connector across the entire J-3 strip (located at the rear left of the drive circuit board). Another difference is in the way the circuit board is laid out. The Tandon drive also has a servo board in the back. This is used for adjusting drive speed. The

FIG. 4-1 A Tandon drive. Note servo board on the back end.

Williams: Repair & Maintenance of Your IBM PC (Chilton)

THE DISK DRIVES

FIG. 4–2 A CDC drive.

CDC drive does not have a servo board. Speed adjustment for a CDC is made on the top circuit board.

If something is wrong with the drives, follow the steps described below and the accompanying photos and drawings.

CHECK THE OBVIOUS

Serious malfunctions in the drive are relatively rare. Most of the problems are brought on by small things, and often by things that have nothing to do with drive operation directly. Don't yank out the drives and tear them apart until you've eliminated all the easy things.

The drive door is made of plastic. It has the unfortunate tendency to break inside. Even if the break hasn't completely separated the door from the drive, it could still prevent the drive from functioning properly. The hub may not make a secure contact with the diskette, or the drive may simply "think" that the door hasn't been closed.

As the drive door goes down, the diskette is placed in the correct position for the spindle to make a positive contact and for the read/write heads to be close enough to do their job. If the door doesn't come down and lock into place properly, the computer is likely to see the drive as having an open door. The diskette can't spin, the program can't load. The usual result of a broken door in drive A is that the computer will load in "Cassette BASIC." The problem could also be intermittent, with data being read or recorded with errors. Drive door problems will also often make a double-sided drive think it is single-sided.

Replacement of the door is simple. The door assembly is held in place by two screws located at the top front of the drive (one screw in a CDC drive). Push down on the metal lever arm and remove the screws (Figure 4-3). The assembly should lift out easily, and the replacement goes back in just as easily. Alignment is important but not critical. Before you tighten the screws make sure that the door closes securely and that the front of the door is flush with the drive bezel (Figure 4-4).

One of the biggest problems with the drive isn't the drive at all but the software. If the software isn't operating the way it should, it could seem that the drive is malfunctioning. (See Chapter 3.)

If you haven't cleaned the heads in some time, there could be deposits that are preventing the heads from operating. False or intermittent data read/write could mean that a dirty head is causing a problem. (See Chapter 7 for more information on cleaning the drive heads.)

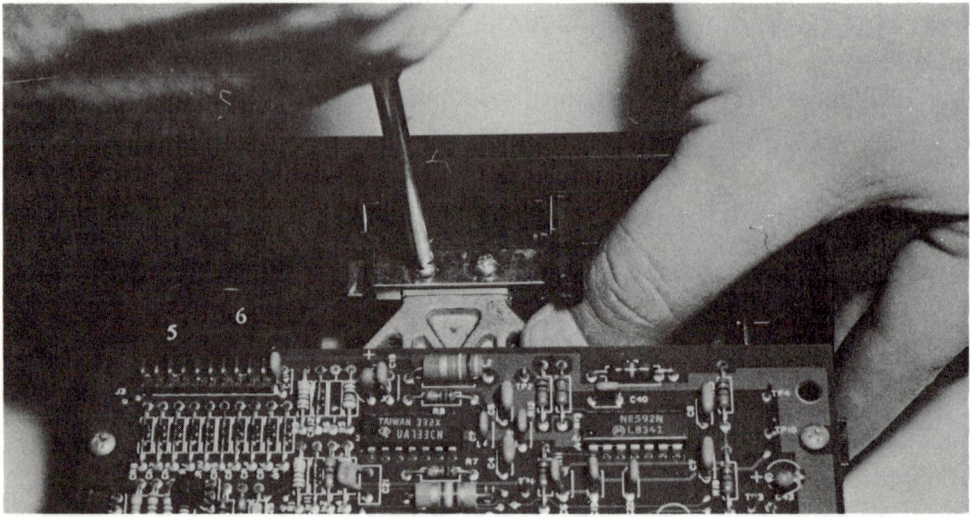

FIG. 4-3 Push down on lever arm, then remove screws.

Williams: Repair & Maintenance of Your IBM PC (Chilton)

THE DISK DRIVES

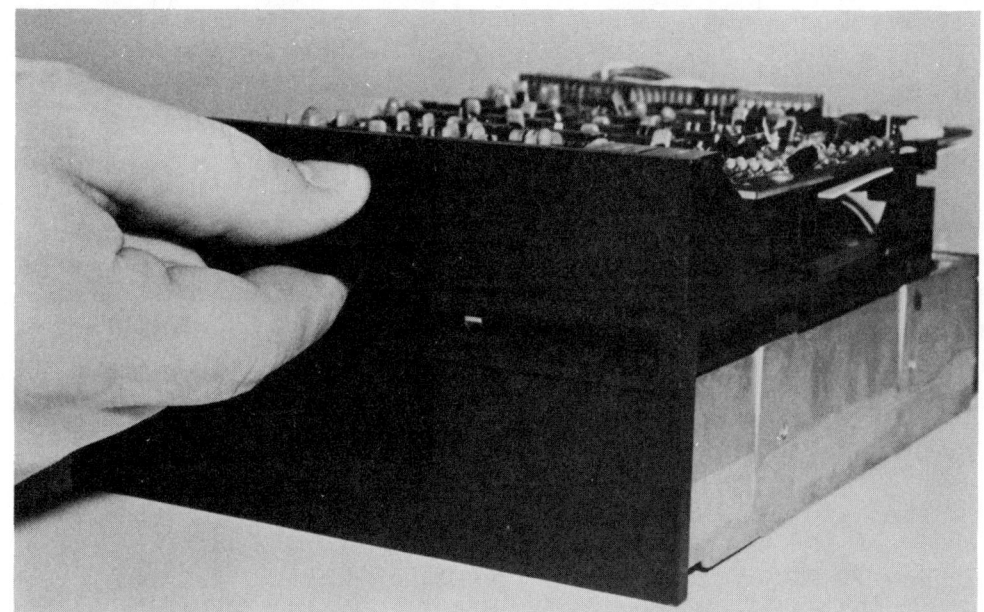

FIG. 4–4 Replace door and check for alignment to be sure the door closes properly.

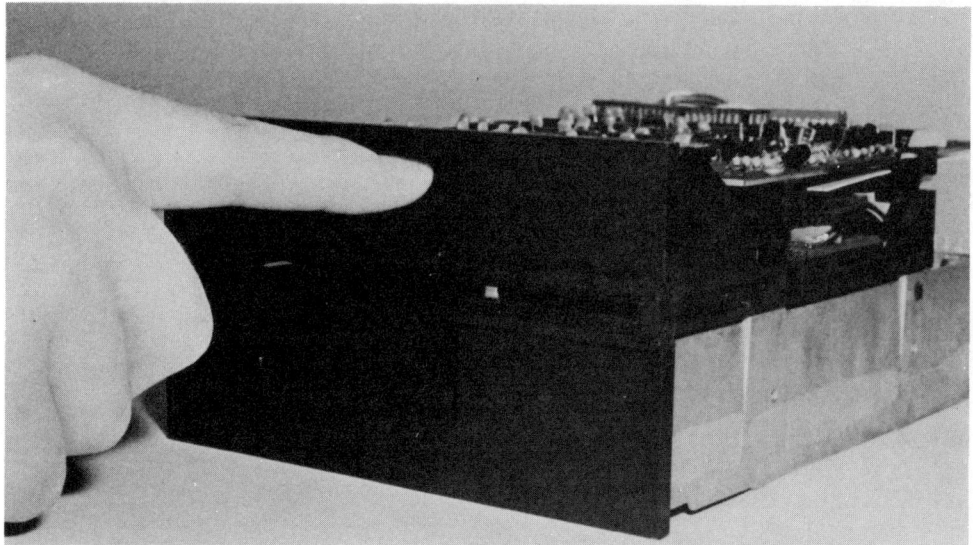

FIG. 4–5 Always handle the drive door with care. One way to safely open the door is to press the top with your finger. The door can't slam open because your finger is in the way.

Have you made any changes in your system? If so, you may have changed the switch settings inside the computer. (See "Helpful Tables and Charts" at the end of this book or your *Guide to Operations* manual for correct switch settings.) If everything was working perfectly before the change, you've done something wrong in making the change.

Are the cables secure? Try unplugging them and pushing them back into place. If the contacts appear to be a little dirty, clean them. You can use either a cleaner that doesn't leave a residue or the eraser of a pencil (taking care to keep the eraser shreds out of the computer). Be sure to shut down power before removing or inserting any cables or circuit boards. Failure to do so will almost certainly damage the board and the computer.

DIAGNOSTIC STEPS

Before you go into the actual diagnostics to find out what has gone wrong with a drive, you can try several easy things.

The first is to run the diagnostics diskette. This will check the drive in its various functions. (This assumes that drive A will operate and load the disk-

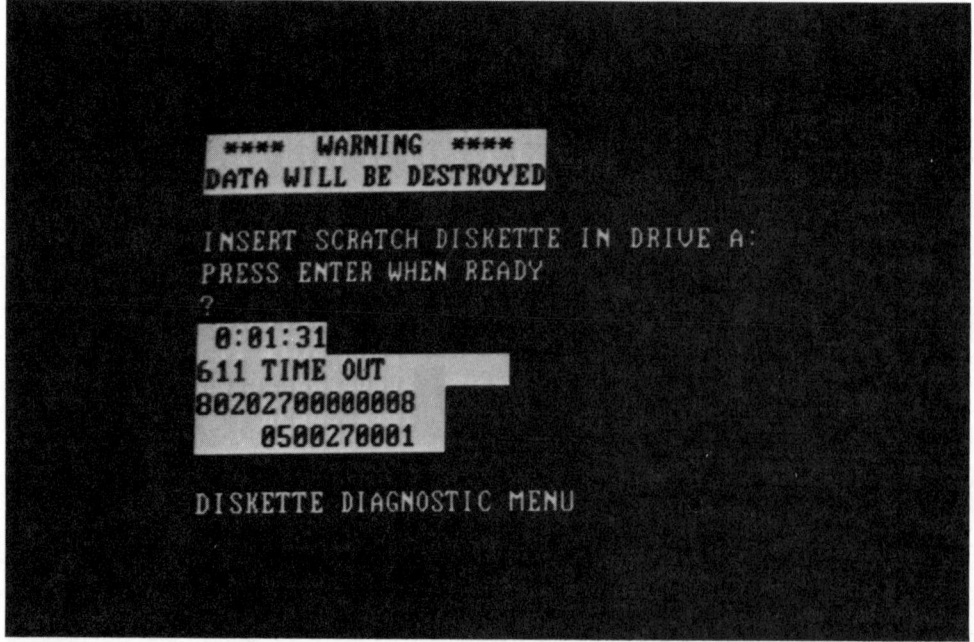

FIG. 4–6 Typical "Time Out" error code. The fourth digit in the code is 0, which means that the problem is with drive A.

Williams: Repair & Maintenance of Your IBM PC (Chilton)

THE DISK DRIVES

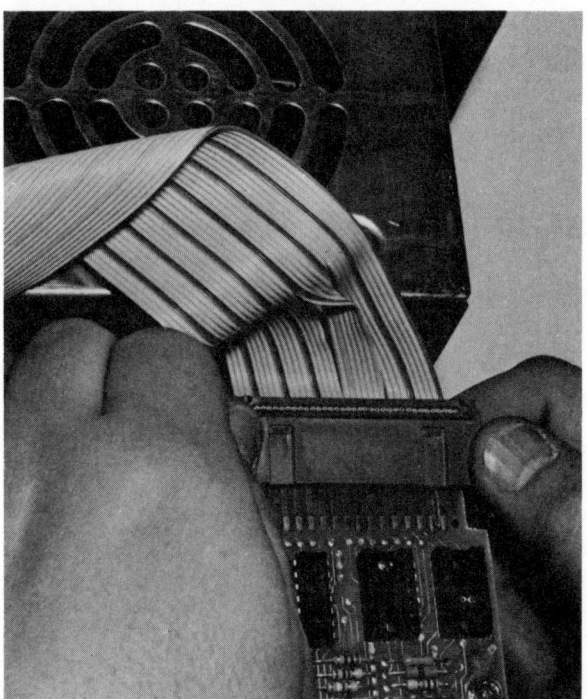

FIG. 4–7 Swapping the drive connectors should give you a different error code. Note the twisted pair. Drive B is now drive A to the computer. Be sure to shut off power before making the swap. If the same problem occurs after the swap, the trouble is elsewhere in the computer.

ette.) The diagnostics may also give you a "Time Out" code (Figure 4–6). If it does, look at the fourth digit in the series. A 0 tells you that the problem is in drive A. A 1 shows the error to be in drive B.

Next you can try a simple swap. (Run diagnostics again after the swap to see if the code changes.) Open the cabinet and change the connections to the drives so that drive B is connected as drive A and vice versa (Figure 4–7). After doing this, you'll know which drive is at fault. If the symptoms are the same, chances are that the fault is somewhere other than the drives.

See also Chapter 6, "Power Supply," since a faulty drive *could* keep the entire system dead. (If the fan is operating, power *is* getting to the computer.) Try using each drive as drive A, with the drive B connector eliminated. Disconnect the cables from drive B and run the diagnostics. Then switch the lines so that the second drive acts as drive A, with the first drive disconnected. Run diagnostics again and watch for a change in the operation, in the error code, or in both.

If you need to disassemble the drive, see the instructions and illustrations beginning on page 87.

For all following steps, shut down the power and wait five seconds before flipping the switch on again. This resets the built-in protective circuitry and allows you to make accurate testings. It is also suggested that you perform each step at least twice, since the probes of your meter may not have been touching the right spots.

It is extremely important that you not create an accidental short by touching the wrong pins and test points. If your hands are shaky or you have any doubts, let a professional take care of the job. You can do a lot of damage by being careless.

FIG. 4—8 Locations of P9-1 and P9-2 on Tandon drive.

Williams: Repair & Maintenance of Your IBM PC (Chilton)

THE DISK DRIVES

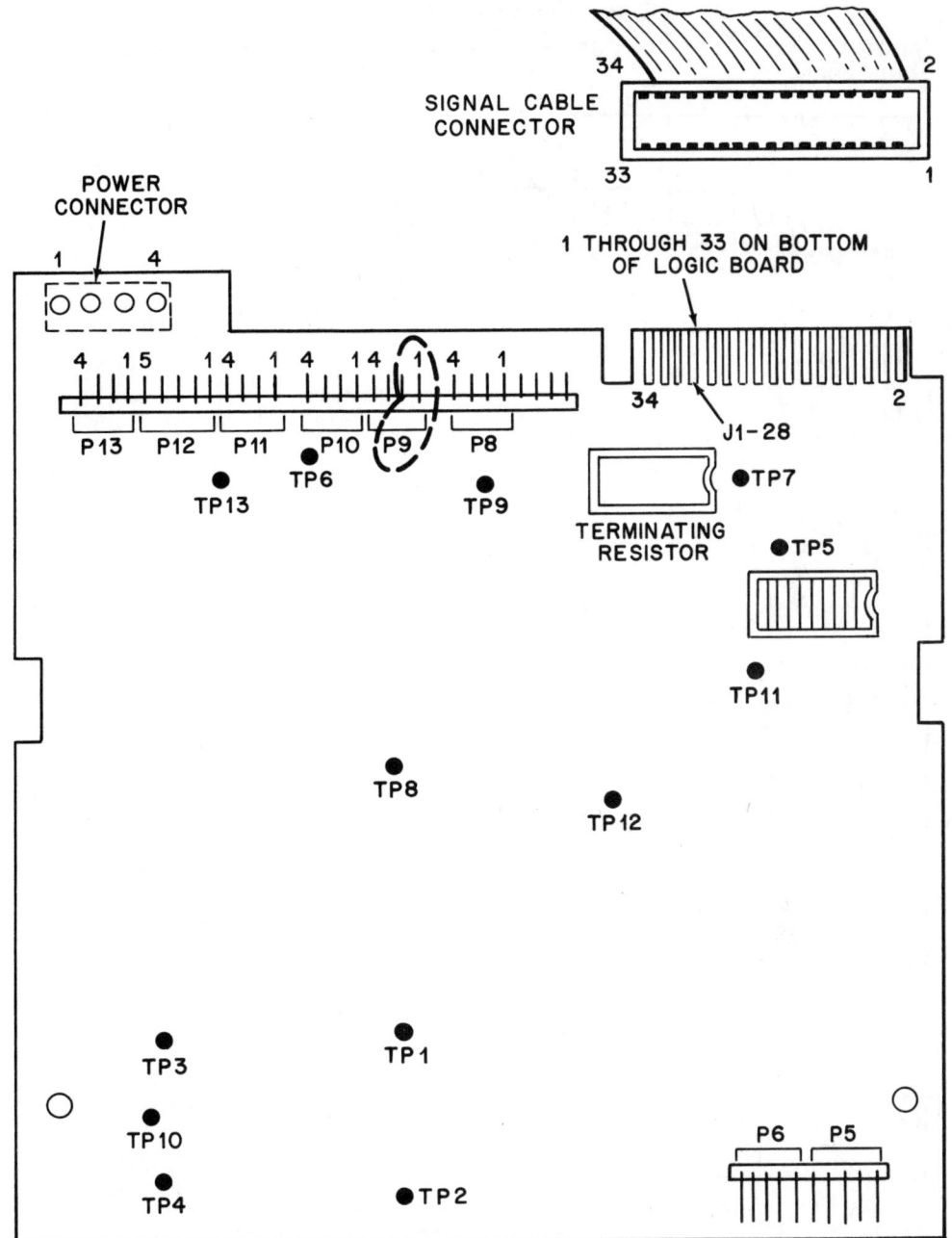

FIG. 4–9 Locations of P9-1 and P9-2 on Tandon drive.

Williams: Repair & Maintenance of Your IBM PC (Chilton)

THE DISK DRIVES

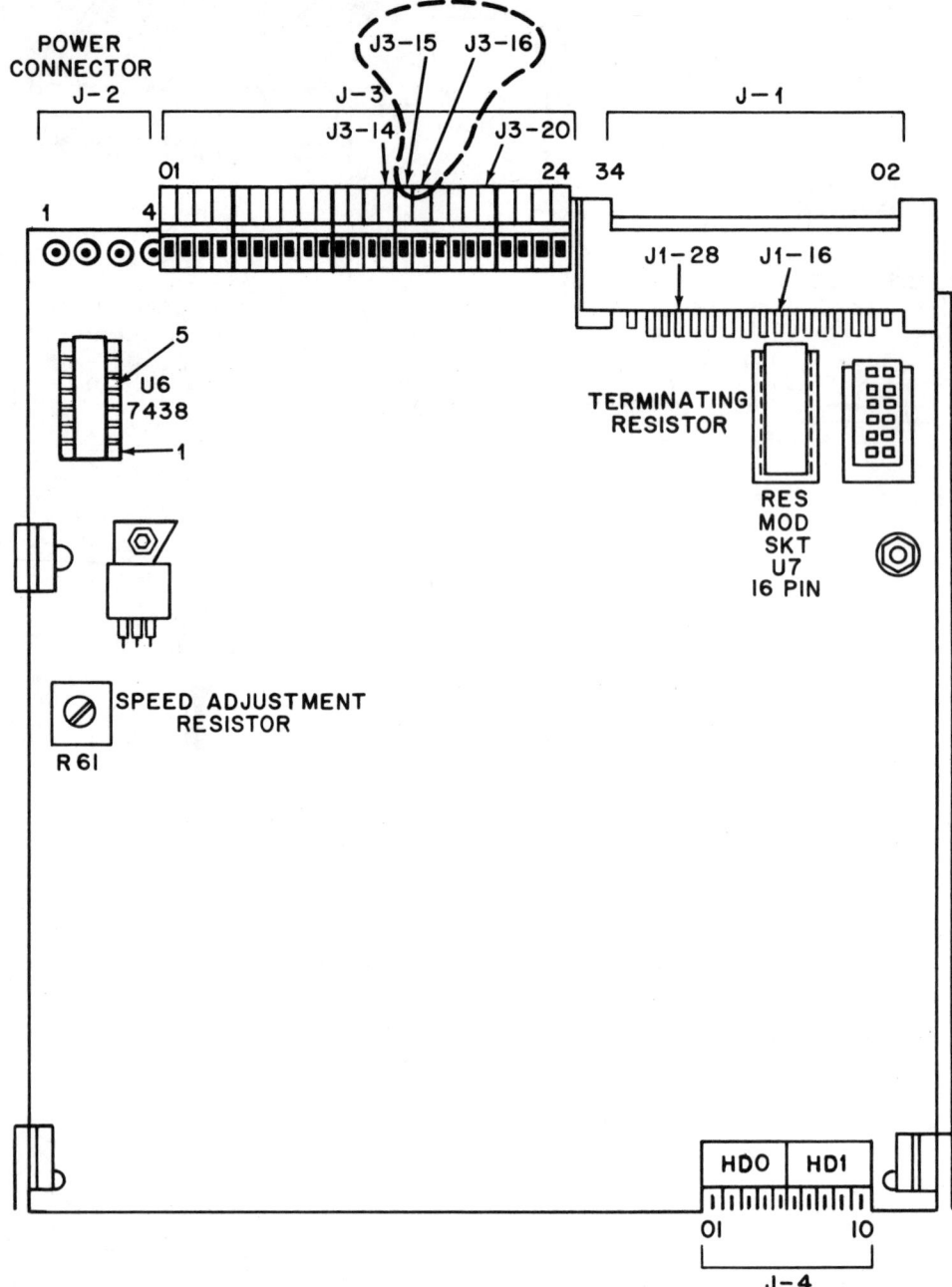

FIG. 4—10 Locations of J3-15 and J3-16 on CDC drive.

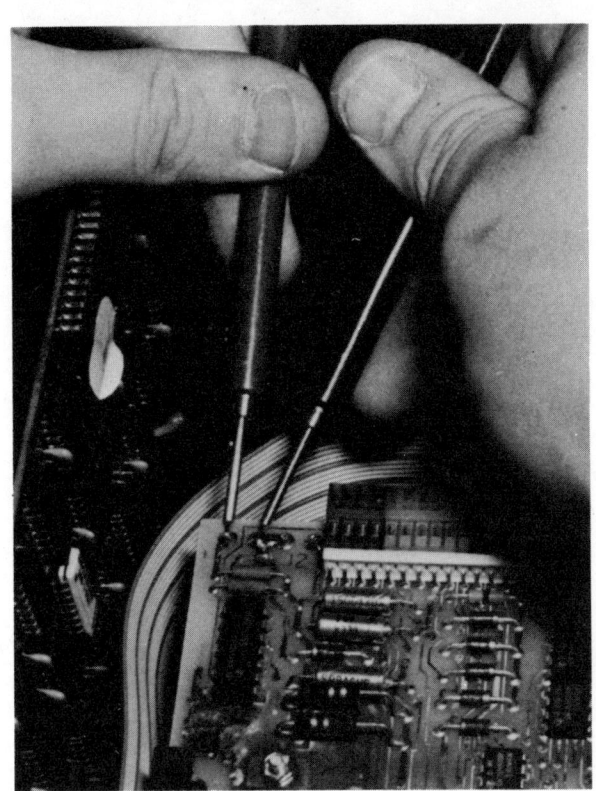

FIG. 4–11 Checking for power to the drive.

STEP 1

The normal response for a properly working drive is for the LED (the little red light) to come on before the beep. If the light does not come on, there is an easy way to tell if the problem is in the LED or in the power supply. (If it lights up, skip to Step 5.)

Get out your meter and set it to read 1.5 volts dc. The common (ground) probe will touch P9-1 on Type 1 drives or J3-15 on Type 2 drives. The other probe should touch P9-2 or J3-16 depending on drive type. (See Figures 4-8, 4-9, and 4-10). Start the computer again and watch the reading you get. It should hit 1.5 volts just before the beep sounds and whenever the spindle is turning in the drive. If the reading is correct but the LED still does not light, the LED needs to be replaced.

To do this, remove the drive from the computer, and then take off the front panel. There are two screws on the bottom and two retaining bushings on top

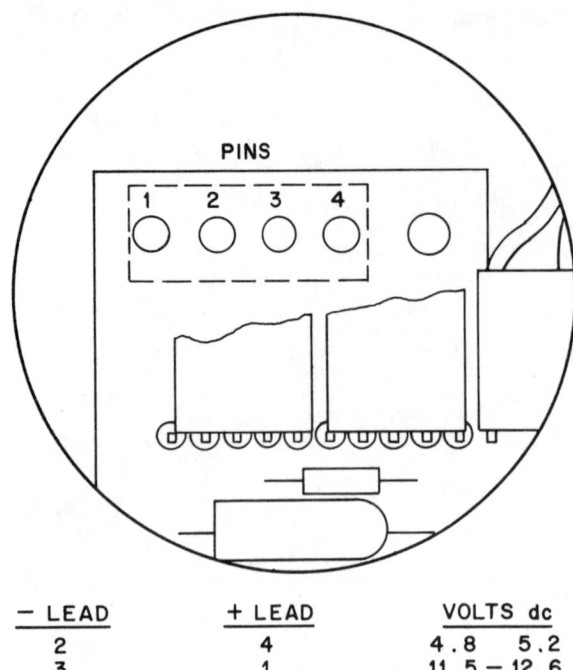

PINS

− LEAD	+ LEAD	VOLTS dc
2	4	4.8 5.2
3	1	11.5 − 12.6

FIG. 4–12 Pin locations and voltages for power check.

FIG. 4–13 Pin 3 on connector 21 on a Tandon drive.

FIG. 4–14 Pin 3 on connector J3 on a CDC drive.

with a Tandon; the CDC has two holding screws on each side. The LED is held in place by a bushing and is connected with a plug. After replacement, the LED goes back in just as easily by reversing the steps.

With no diskette in place, look inside the drive as you turn on the power. At the end of the power-on self test, the spindle should begin spinning (at the same time as the lighting of the LED). If it does you may be able to skip Steps 2 through 9 and go to Step 10 (testing and adjusting the drive speed).

STEP 2

The next step is to check the power going into the drive. On both types the power cable from the drive comes into four pins, numbered from left to right, in the upper left corner of the circuit board (see Figures 4–11 and 4–12). The voltage between pins 2 (ground) and 4 should be between 4.8 and 5.2 volts dc. The voltage between pins 3 (ground) and 1 should be between 11.5 and 12.6 volts dc. If it is not, the fault is in the power supply. (See Chapter 6.)

FIG. 4–15 Tandon terminating resistor location.

If you have Tandon drives, check the voltage coming into the servo board (located at the back of the drive). You should read approximately 12 volts dc between pins 1 and 2 (com) on connector 20. Between pin 4 and ground you should get a reading of about 5 volts dc. If the voltage here is correct and the voltage on the main circuit board on top of the drive is incorrect, you'll know that the problem is in the main circuit board. Otherwise it is in the power supply.

Williams: Repair & Maintenance of Your IBM PC (Chilton)

THE DISK DRIVES

Finally, check the voltage between pin 3 on connector 21 and ground for a reading of between 3 and 9 volts dc when the LED comes on. If you have a CDC drive you'll test for this voltage between pin 3 on the J3 connector and ground. (See Figures 4–13 and 4–14.)

STEP 3

If the readings have been in these ranges, check the terminating resistor. This looks just like an IC (refer to Figures 4–15, 4–9, and 4–10 for location). First check to see if this resistor package is inserted correctly into drive A and has been removed from drive B. Also check to see if both drives have the jumper wire correctly installed (see Chapter 8, "A Second Drive").

To test the drive circuit board, set your meter to read 5 volts dc. The voltage between pin 12 on the signal cable (Figure 4–17) and ground at the beginning of the power-on self test (POST) should be about 5 volts. If this voltage isn't present, the circuit board on the drive is faulty and needs replacement.

Try the same test again, this time watching for a decrease in the voltage from about 5 volts to near 0 volts while POST is taking place. A decrease of 5 to 0 volts should also occur at pin 18 just before the beep at the end of POST. The decrease means that the circuit card is good and that the problem is in the

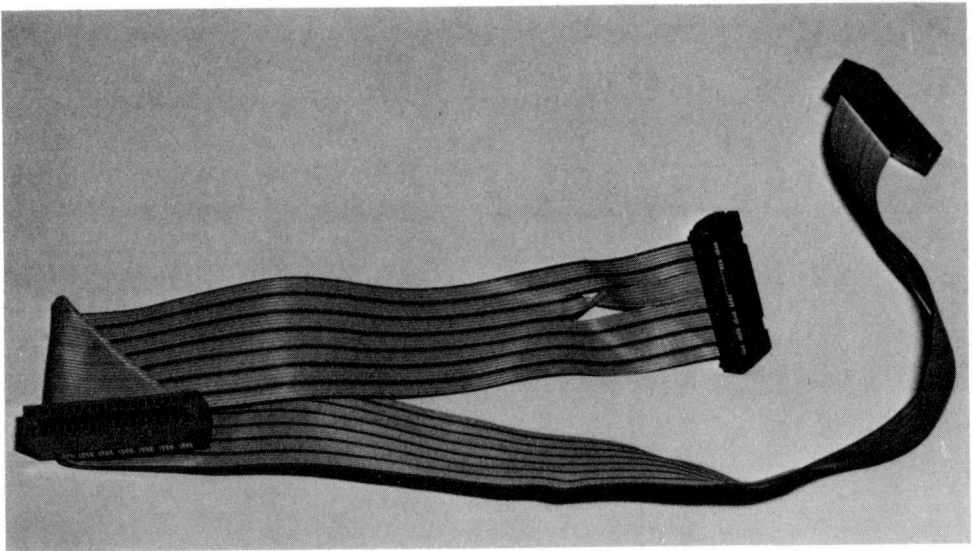

FIG. 4–16 Signal cable for the drives. The end with the twisted wires is attached to drive A.

Williams: Repair & Maintenance of Your IBM PC (Chilton)

mechanical part of the drive. Replacement of the drive assembly will be necessary.

STEP 4

If there was no decrease in voltage during POST, check the signal cable for continuity. After visually checking the cable for obvious damage set your meter to read ohms (x1) and follow the diagram in Figure 4-17 and Table 4-1. If all

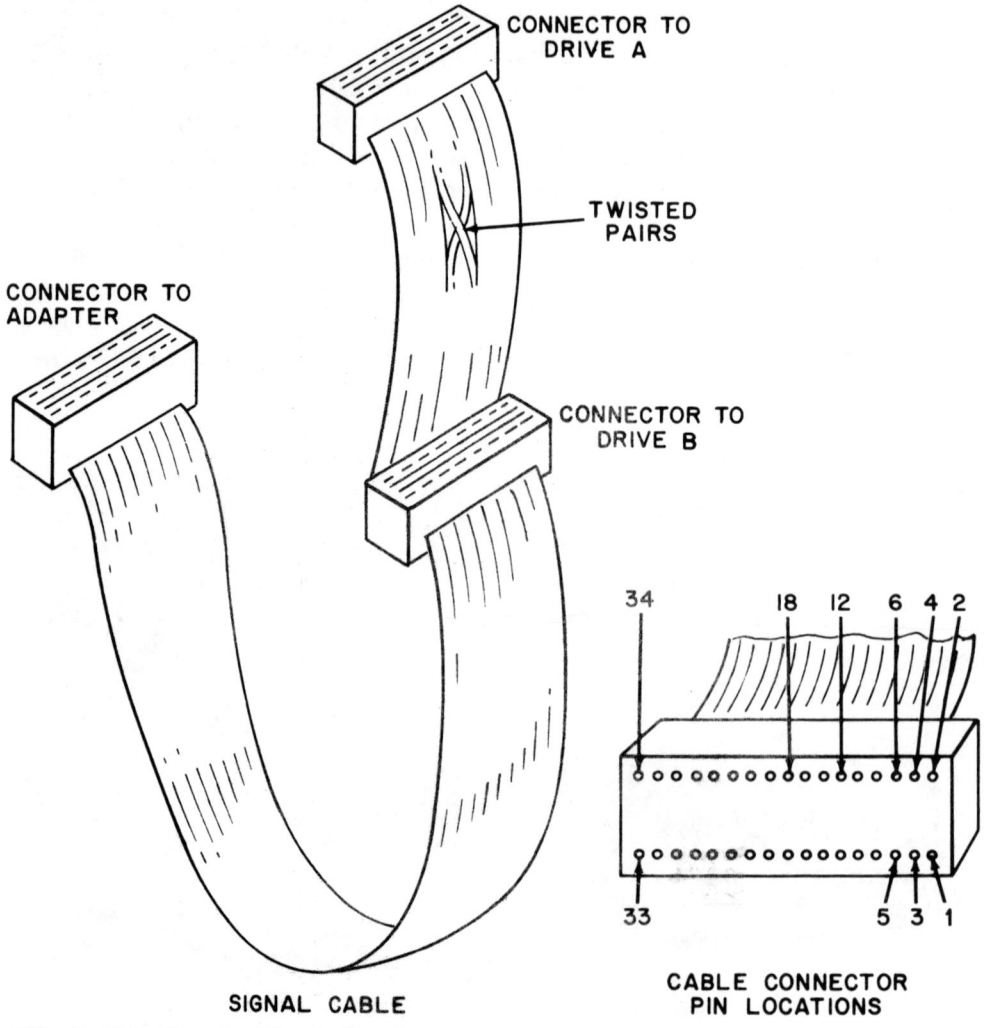

FIG. 4—17 Signal cable pin locations.

Williams: Repair & Maintenance of Your IBM PC (Chilton)

TABLE 4–1
Checking Signal Cable Continuity

All wires run from pin to pin by number. Pin 1 on one side is connected to pin 1 on the other and so on. This is true for all pins *except those going to drive A as listed below.* This is the twist you see in the cable in Figure 4–16.

Drive A	
10	16
11	15
12	14
14	12
15	11
16	10

wires in the cable check out, the problem is in the driver adapter card. If not, replace the cable.

STEP 5

If everything is fine to this point, or if the first tests don't apply (the LED lights correctly), you will be checking for a voltage change across other points.

For a Tandon (Type 1) drive, set the meter to register in the 0.5 volt dc range. (If your meter doesn't have a setting for a half volt, simply use the lowest setting). The test is to watch for a change from 0.5 volts dc to near 0 between P10-2 and ground (see Figure 4–18) as a diskette is inserted into the drive.

For a Type 2 drive (CDC), set your meter to read 5 volts dc. This time the voltage should *increase* from 0 to near 5 volts between J3-20 and ground (see Figure 4–19).

If this doesn't happen, then the drive is faulty and requires replacement. If it does happen, continue to the next step.

STEP 6

To test both drive types, set the meter to read 5 volts. You'll be watching for a decrease in voltage from 5 volts to 0 volts as a diskette is put in. For a Type 1 (Tandon) the probe will touch TP7. For a Type 2 (CDC) it will touch pin U6-5. (See Figures 4–18 and 4–19 for locations.)

If the results don't show the decrease, the circuit board on the drive is at fault. If the results are correct, the drives should be functioning well enough to perform a "format" routine.

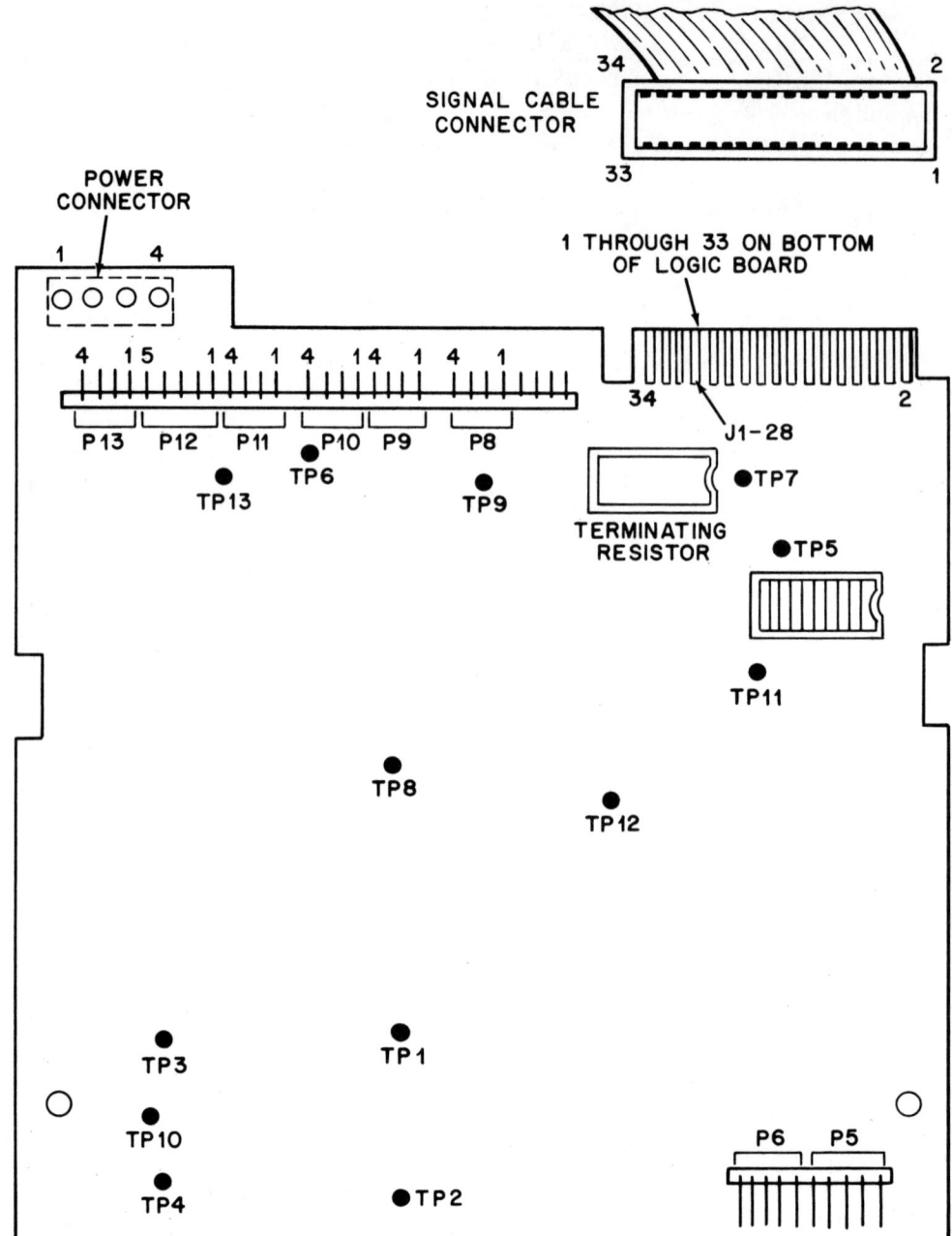

FIG. 4–18 Location of pins and test points on the Tandon drive.

Williams: Repair & Maintenance of Your IBM PC (Chilton)

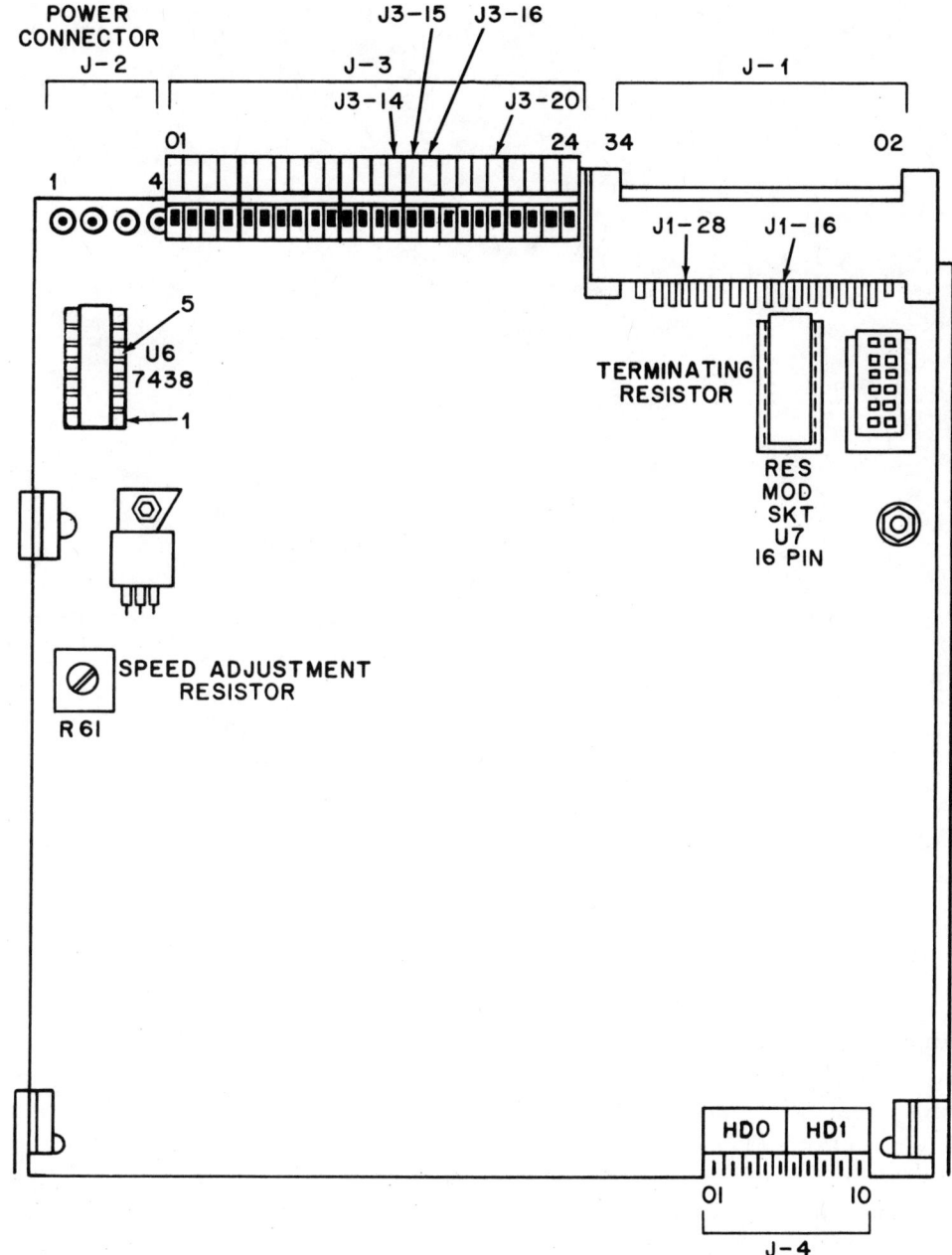

FIG. 4-19 Location of pins and test points on the CDC drive.

Williams: Repair & Maintenance of Your IBM PC (Chilton)

THE DISK DRIVES

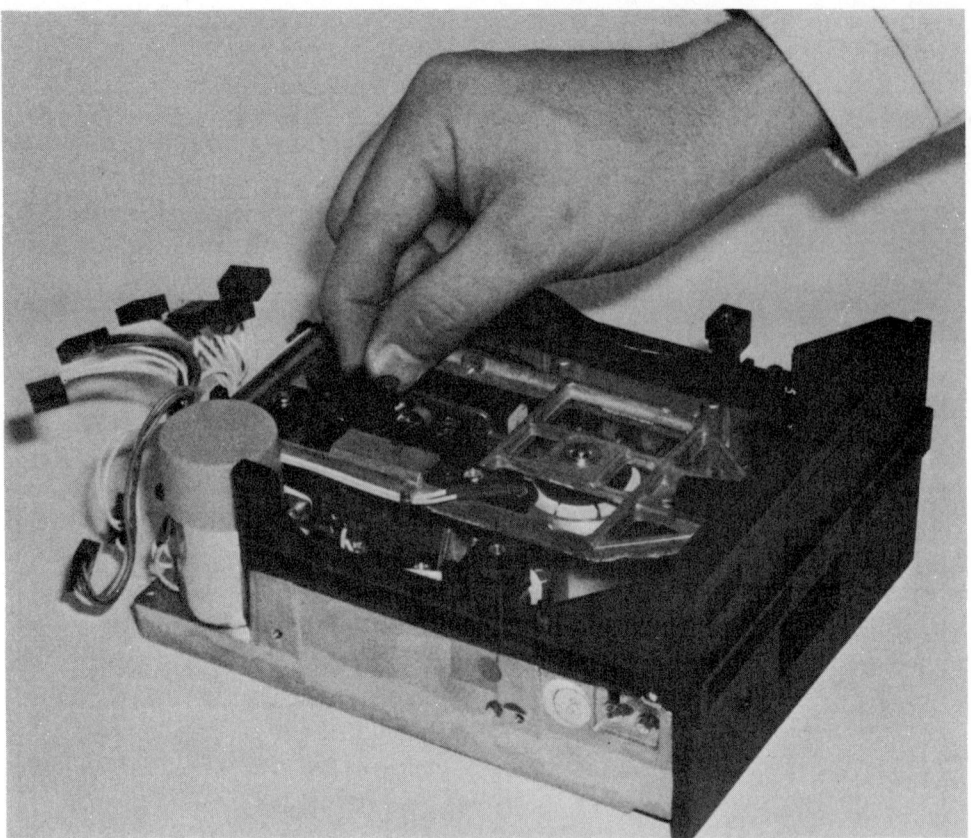

FIG. 4–20 Move heads back and forth to check for sticking or binding. Track 0 is at the back.

STEP 7

Testing for "format" will tell you several things and could save you a lot of grief. The purpose of the next steps is to test the write-protect switch inside the drive to make sure that this switch is operating correctly. If it isn't, a program or data you *think* is protected could be destroyed.

Try to format a diskette that has the write notch covered. It *should not* format, but should display a "Write Protected Diskette" message.

To test the switch electrically, set your meter to the 5 volt dc range. You'll be watching for a change in voltage as you partially insert and remove a diskette.

Williams: Repair & Maintenance of Your IBM PC (Chilton)

THE DISK DRIVES

With a Type 1 (Tandon) drive, there should be an *increase* from 0 to 5 volts between P8-1 and ground. A CDC (Type 2) drive should show a *decrease* from 5 volts to 0 between J3-14 and ground. (See Figures 4–18 and 4–19 for correct locations.) If you do not get these changes in voltage, replace the write-protect switch.

You can also test the switch by checking for continuity across the switch. Disconnect one of the leads to the switch to isolate the switch from the computer. Set the meter to read ohms in x1 range. The reading should go from infinity to near zero ohms as the switch is activated (by inserting a diskette). Perform this test with the power off.

Both types of drive require that you remove the circuit board to replace the switch. The Tandon drive also requires removal of the drive itself. The switch on both is on the left hand side looking toward the drive opening. Mounting screws are: two through the side for a Tandon, and one through the top on the CDC.

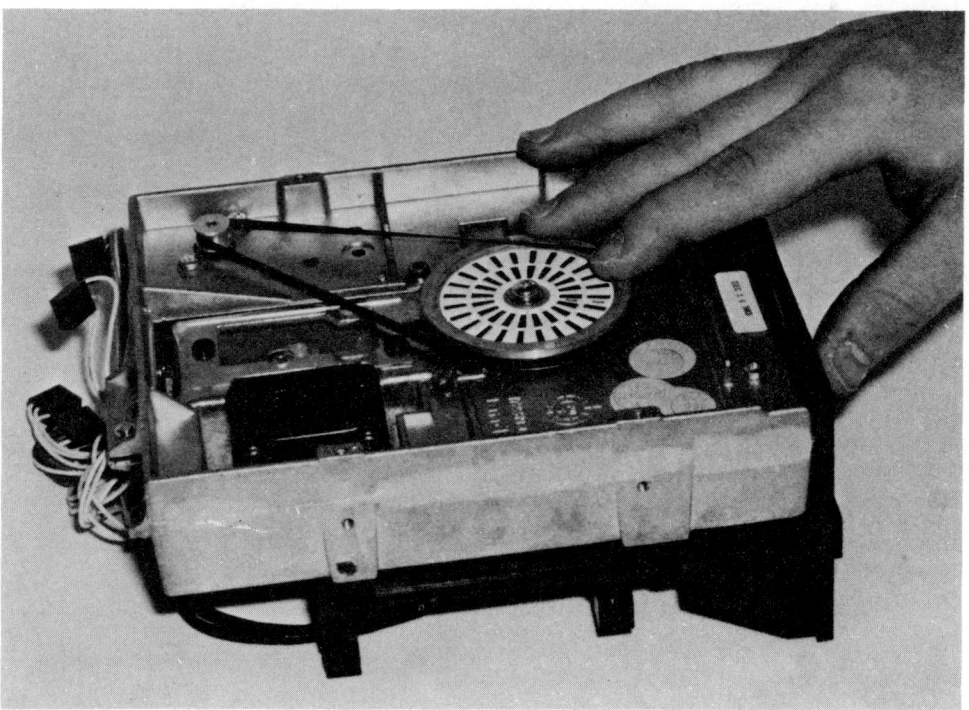

FIG. 4–21 Pulley and belt of the disk drive. Turn gently to check for sticking or binding.

Williams: Repair & Maintenance of Your IBM PC (Chilton)

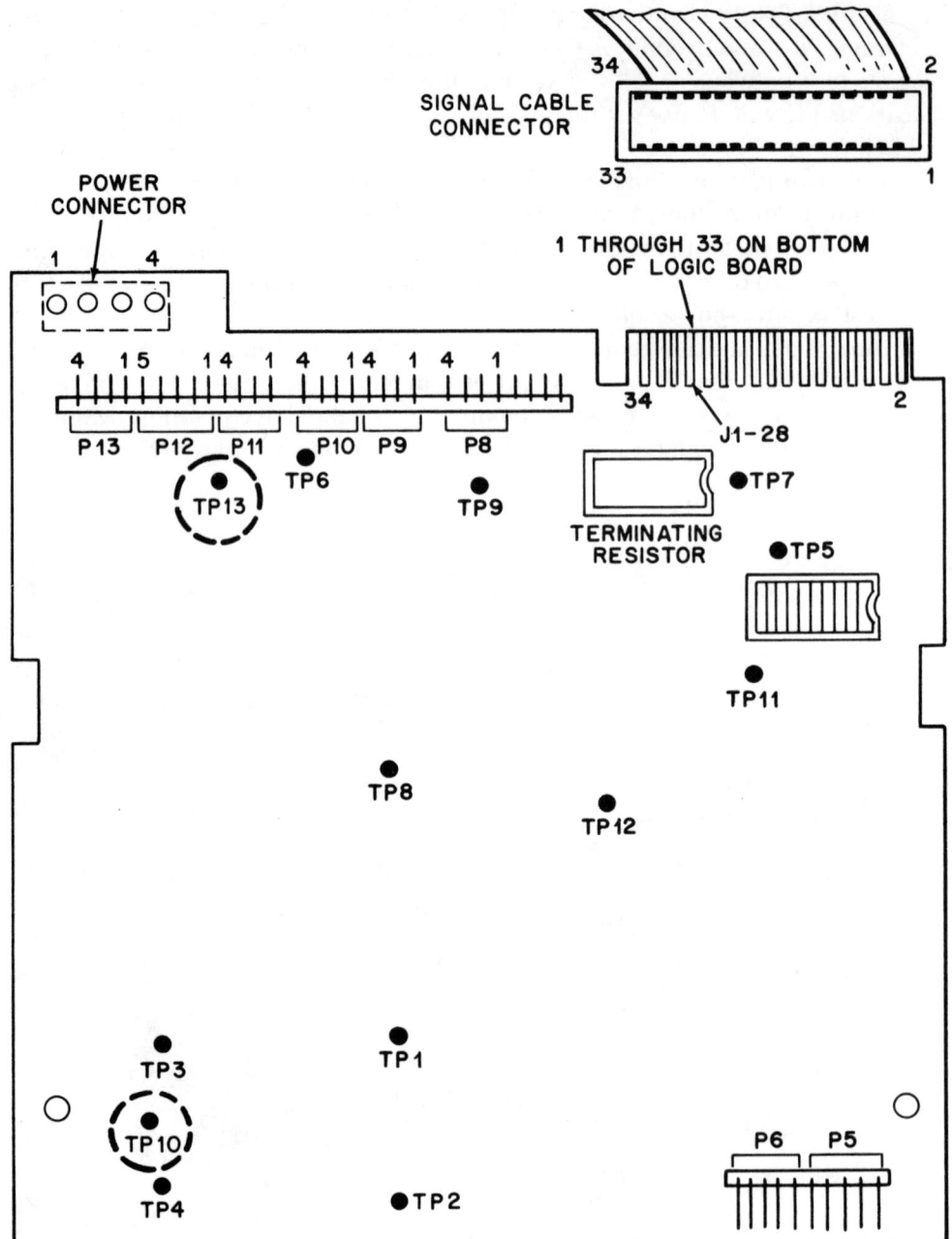

FIG. 4–22 Location of TP10 and TP13 on the Tandon drive.

Williams: Repair & Maintenance of Your IBM PC (Chilton)

THE DISK DRIVES

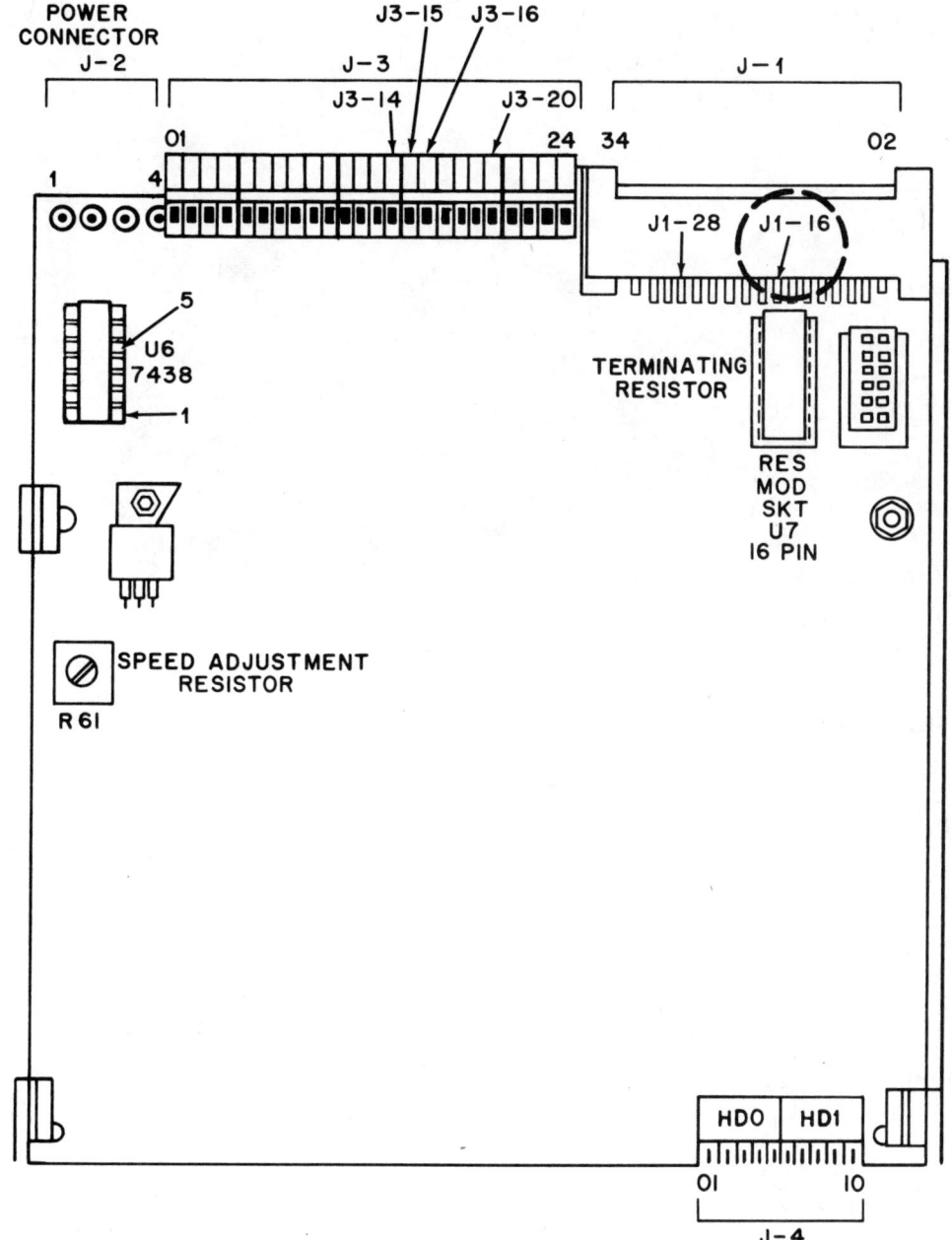

FIG. 4–23 Location of pin J1-16 on the CDC drive.

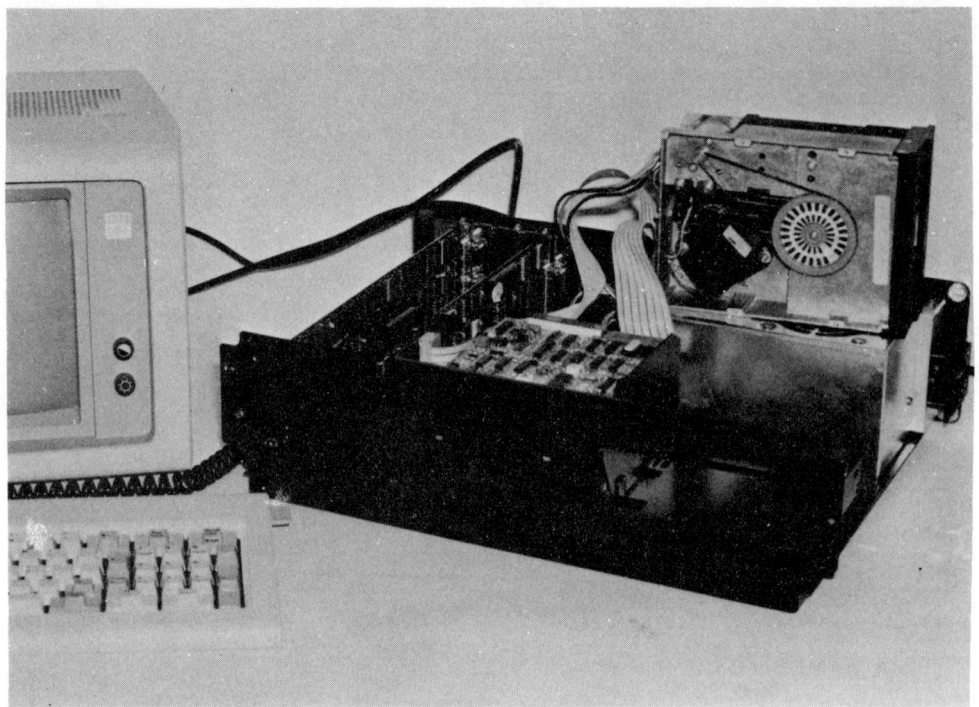

FIG. 4–24 Setting up to test and adjust the drive speed.

STEP 8

If the write-protect switch passes the test above, check the voltage between pin J1-28 and ground (see Figures 4-18 and 4-19). As you insert and remove a diskette, the write-protect switch is activated. The beginning voltage with the LED lit should be about 5 volts. (If it isn't, go back to Step 4.) As you insert the diskette and activate the write-protect switch, the voltage should drop to zero. If it does not decrease to nearly zero, the trouble is in the drive circuit board. If it does decrease and the drive is definitely causing problems, the adapter card is causing the trouble.

STEP 9

Remove the suspected drive from the computer and take off the circuit board. This will allow you to see the read/write heads. Gently move them (front to back) with your hand (Figure 4–20). There should be a small amount of resis-

Williams: Repair & Maintenance of Your IBM PC (Chilton)

THE DISK DRIVES

tance to the motion, but the movement should be possible without any sticking or binding.

Turn the drive over. You'll see a pulley and belt. Turn the pulley by hand to check for sticking or binding (Figure 4–21). Look at the belt to make sure that it is securely in place.

If both things seem okay, partially attach the circuit board again, leaving it just loose enough so that you can lift it to watch the heads. Reattach all connectors except P5 and P6 (HD0 and HD1 on a CDC drive). See Figures 4-18 and 4-19 for locations. Push the heads away from the rear of the drive. This is track 0 (with track 39 being at the front of the drive). Turn on the power. During the self test the head should move to track zero and away from it again. If it does not do this, the drive assembly is bad and will have to be replaced.

Shut off the power and connect the drive completely again. Set the meter

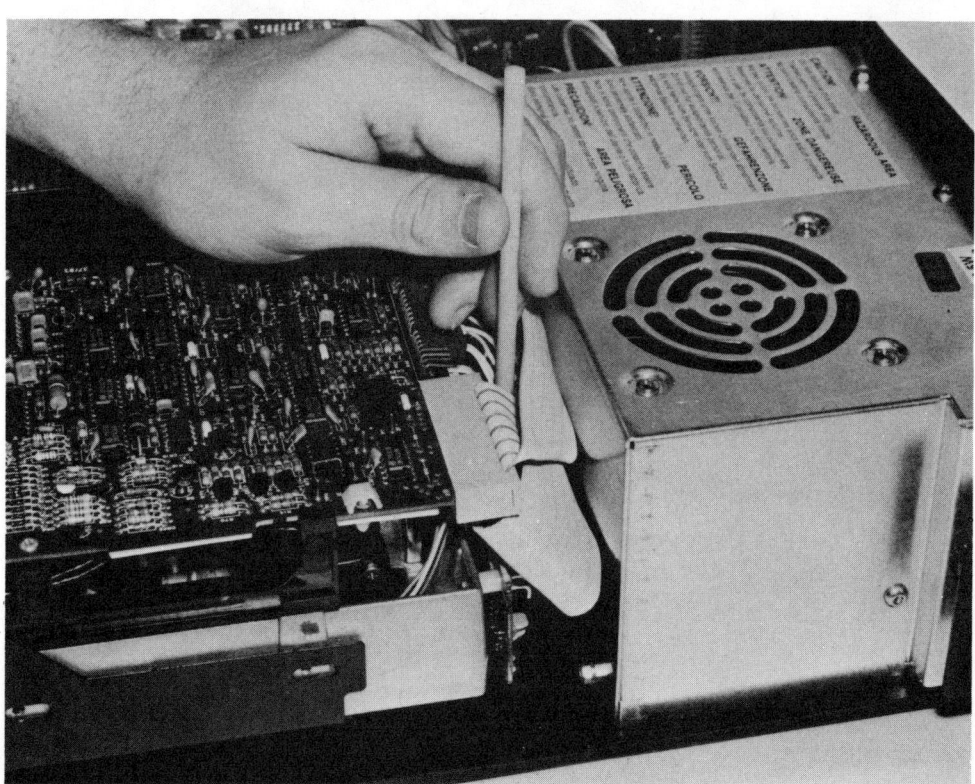

FIG. 4–25 Speed adjustment control on the Tandon drive.

Williams: Repair & Maintenance of Your IBM PC (Chilton)

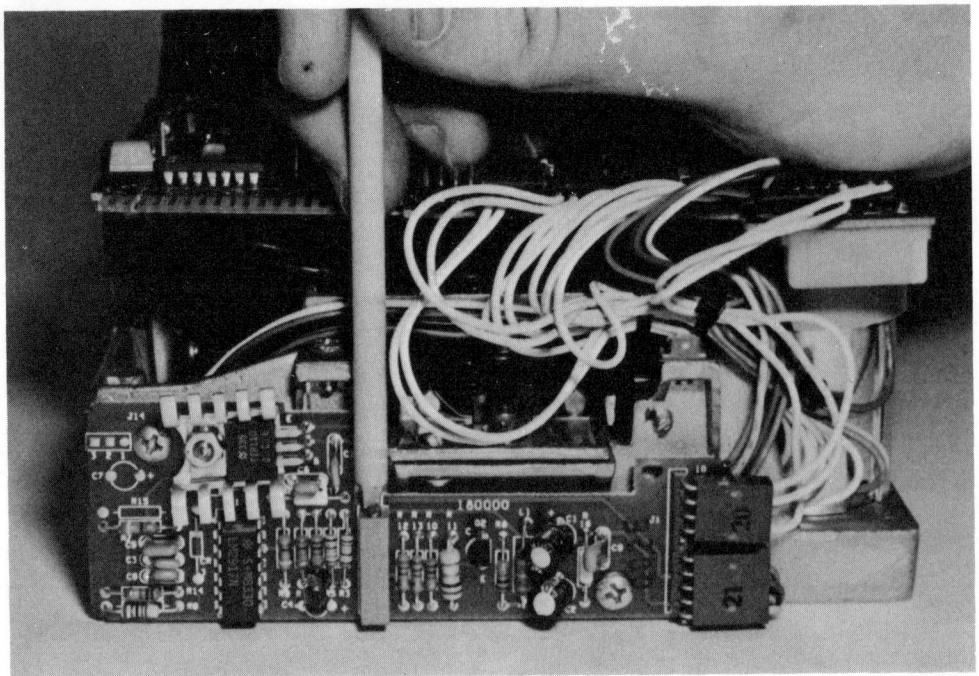

FIG. 4–26 Close-up of speed adjustment control on the Tandon drive.

to read in the lowest range. For a Tandon drive you'll be testing the voltage between TP5 on the circuit board (right rear) and ground (Figure 4-18). For a CDC drive the test is between pin 1 on U6 (an IC socket at the left rear) and ground (Figure 4-19). Apply power. As the LED comes on you should see an increase in the voltage of about 0.2 volts dc.

SPEED ADJUSTMENT

There are several ways to test the speed of a drive. Although the standard diagnostics diskette will not do this, the advanced diagnostics diskette will. Verbatim Corp. sells a disk drive analyzer program that tests drive speed and other critical drive operations. If you don't have these things you can still test the drive speed by using a fluorescent light.

Using the advanced diagnostics diskette allows you to make speed adjustments with the drive running. The Verbatim disk will require some experimentation; that is, you must turn the adjustment in each direction a little at a time and run the test again until you have the speed adjusted correctly.

Williams: Repair & Maintenance of Your IBM PC (Chilton)

THE DISK DRIVES

To test and adjust drive speed using a fluorescent light, remove the drive and tip it on its side. For a Tandon drive, connect a jumper between TP10 and TP13 on the circuit board (Figure 4–22). For a CDC drive, connect the jumper between pin 16 on the J-1 connector strip (right rear) and ground (Figure 4–23). When you apply power, the jumpers will keep the drive motors going while you make the adjustments.

Insert a scratch (blank) diskette in the drive and turn on the power. By watching the marks on the pulley under the fluorescent light, you should be able to tell which way to turn the variable resistor for correct speed adjustment. If the drive is operating at the correct speed, the marks on the outer ring should seem to stand still. (The inner marks are for 50 hertz power.)

Whichever method you use, if the speed is off, adjustment is made by turning a variable resistor. In the Tandon drive, the resistor is in an extended "stack" on the servo board at the rear of the drive. In the CDC drive, the adjustment is on top of the circuit board on the left side and near the center. (See Figures 4–24 through 4–28).

(When adjusting drive speed, be very careful not to cause any short circuits.)

DISK DRIVE ANALYZER

As mentioned above, Verbatim Corp. sells a "Disk Drive Analyzer" under its Data Encore subsidiary. The cost is just $40, which is a small investment considering what the program does. The program tests for drive speed, head alignment, spindle clamping, and read/write. After each test you will be told if the drive is good, fair, or poor.

Although you won't be able to make any adjustments with the program, it can keep you informed as to the condition of your drives. A regular testing (which takes about 2 or 3 minutes) will alert you if the heads are going out of alignment or if something else is about to destroy the data. The drives are too critical a part of the computer system to let them just run until they fail.

SOME OTHER PRECAUTIONS

You can avoid many drive problems simply by prevention. Keep the environment as clean as possible. Dust and other contaminants can create havoc with the drive.

Regular cleaning of the read/write heads is a good practice. How often you do this will depend on your surroundings and how much the computer is used. *Don't skimp* when you are buying a head cleaning kit. Get the best possible.

FIG. 4–27 Speed adjustment control on the CDC drive.

(More information on regular cleaning and maintenance is contained in Chapter 7.)

If your drive has to be realigned, or if the drive speed is to be adjusted, make new copies of everything that has been recorded in that drive. Use that drive as the "source," with a drive that you know is good for the "target."

The reason for this is that the faulty drive has recorded the data according to its maladjusted characteristics. If you try to read that data on a properly operating drive, you'll probably get nothing but garbage. By using the faulty drive as the "source," you're allowing it to read the data with the characteristics embedded. The good "target" drive will record the data the way it should be.

The Verbatim disk drive analyzer suggests running the test program on a regular basis. With or without this program you should test the drives occasionally. They are a critical part of the computer system. Realizing that the drive is going bad while it's recording some important data is too late.

Williams: Repair & Maintenance of Your IBM PC (Chilton)

THE DISK DRIVES

DISASSEMBLY OF THE DRIVE

Finding parts for the drive can be difficult. If your dealer won't sell them to you, it's possible that you'll have to be satisfied with reducing repair costs simply by being able to tell them what is wrong with the drive, and with knowing what makes the drive work.

If you *can* get the parts, however, repair is generally a simple thing. Disassembly can be done with nothing but a screwdriver. Proceed carefully and be careful not to damage any parts. Even with the photos, take notes as you go.

SUMMARY

The disk drives are probably the most critical part of your computer system. Unfortunately, they are also the devices most likely to cause trouble. This is

FIG. 4–28 Close-up of speed adjustment control on the CDC drive.

Williams: Repair & Maintenance of Your IBM PC (Chilton)

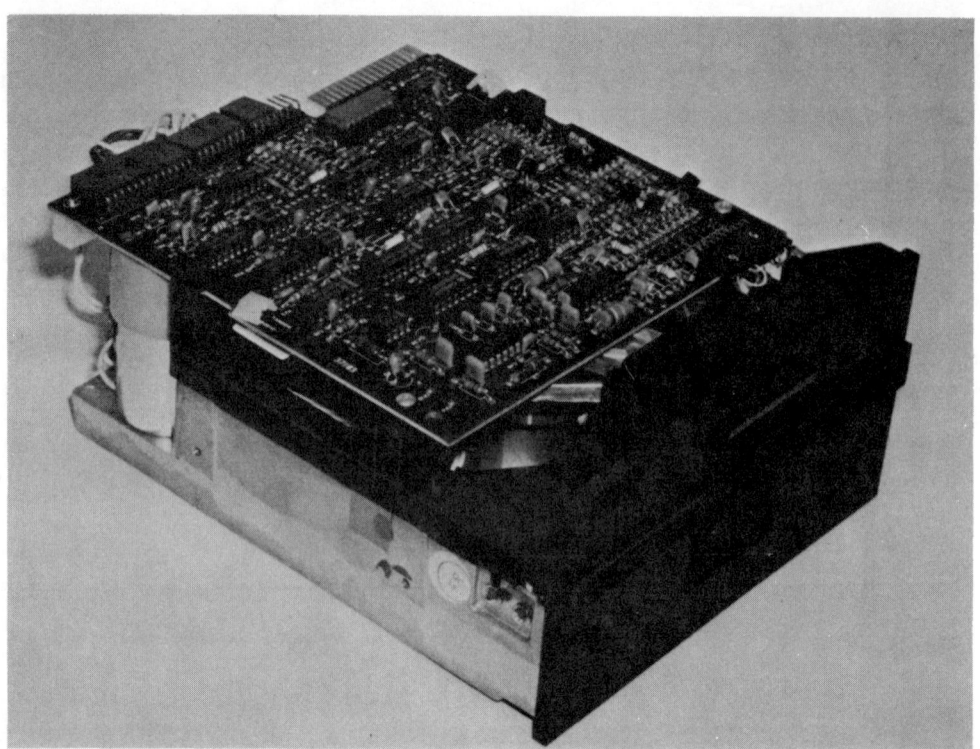

FIG. 4–29 Tandon drive.

FIG. 4–30 Under side of the Tandon drive.

Williams: Repair & Maintenance of Your IBM PC (Chilton)

THE DISK DRIVES

FIG. 4–31 CDC drive.

FIG. 4–32 Under side of the CDC drive.

Williams: Repair & Maintenance of Your IBM PC (Chilton)

THE DISK DRIVES

because they are one of the few mechanical parts in the system. (The only other mechanical device in a standard PC is the cooling fan.)

If the drive malfunctions, existing programs may not operate. Data recorded on a misaligned or otherwise maladjusted drive can disappear once the drive is put back into shape again.

Only rarely will a drive suddenly fail. Most of the time it will give you warning symptoms, such as a faulty read or a faulty write. Both can be caused by other things, but both also indicate that it is time to check out the drives.

Make drive checks and drive maintenance a part of your regular schedule. Clean the heads occasionally. Run diagnostics on them, with either the standard diagnostics diskette, the advanced diagnostics diskette, or one of the commercially available disk drive analyzer programs.

Preventive maintenance is the best possible means of assuring that the drives will give you no problems. Back-up copies of programs and data diskettes made while the drives are operating properly will help to ensure that a drive malfunction is not a disaster.

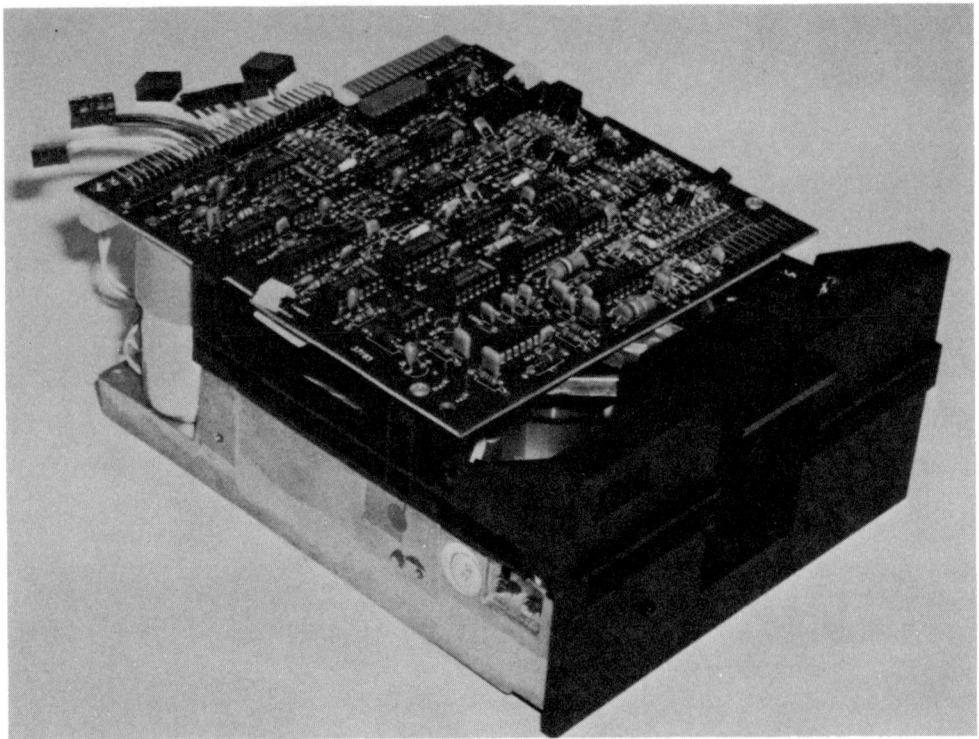

FIG. 4–33   Step 1: Disconnect all cables, *carefully*.

Williams: Repair & Maintenance of Your IBM PC (Chilton)

THE DISK DRIVES

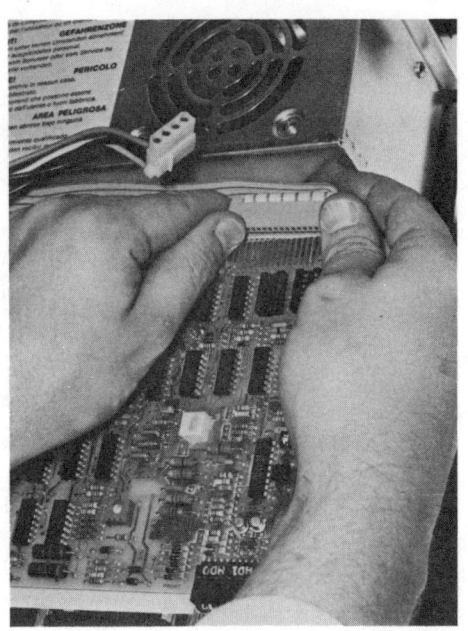

FIG. 4–34 Disconnecting the power cable and signal cable from the drive.

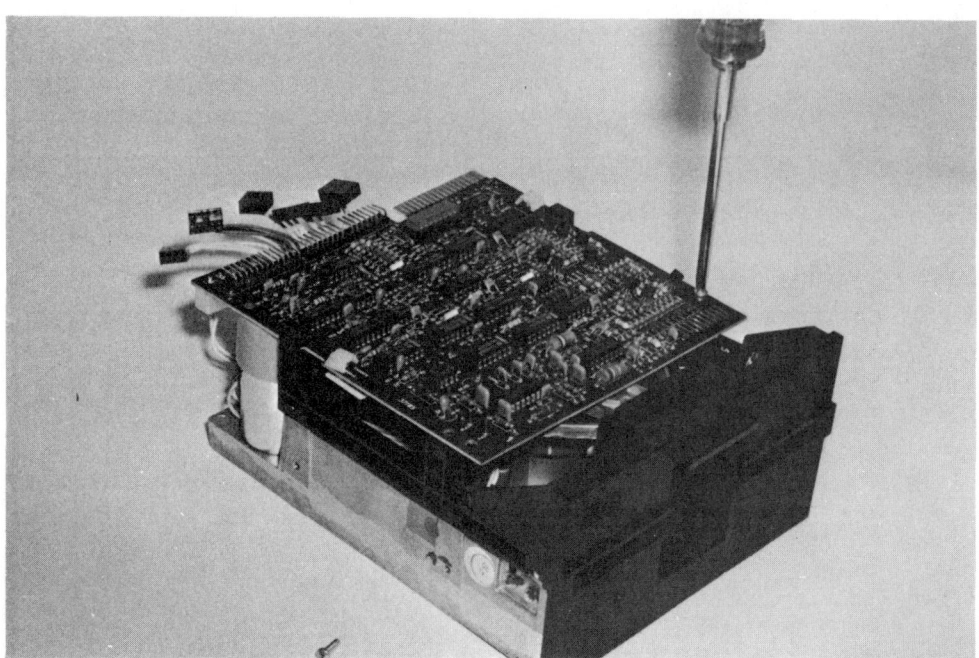

FIG. 4–35 Step 2: Carefully remove the screws from the driver board. (Some drives have Phillips screws; others have hex heads.)

Williams: Repair & Maintenance of Your IBM PC (Chilton)

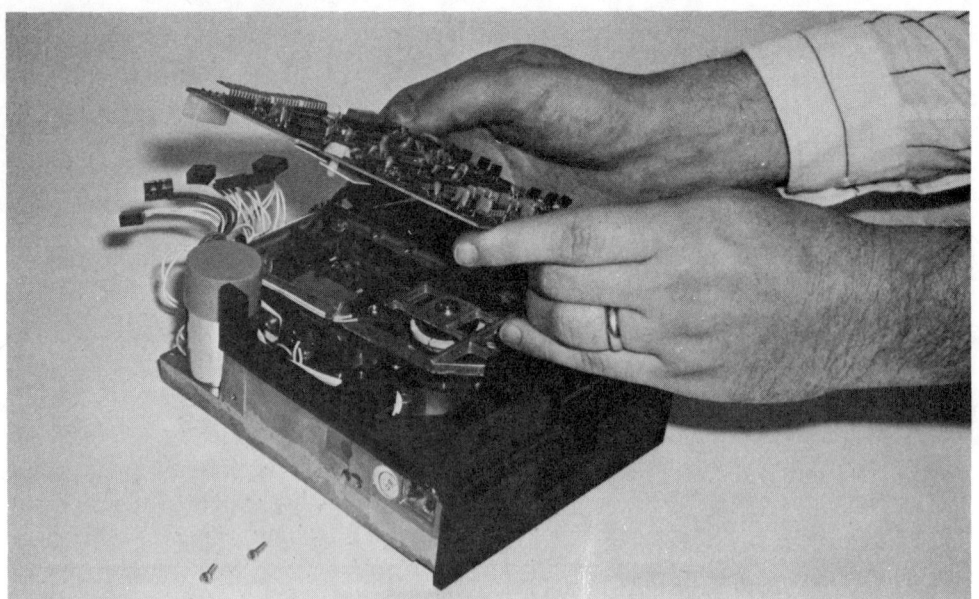

FIG. 4–36 Step 3: Carefully lift away the driver board.

FIG. 4–37 Step 4: Remove shield if your drive has one.

Williams: Repair & Maintenance of Your IBM PC (Chilton)

THE DISK DRIVES

FIG. 4–38   The disassembled drive.

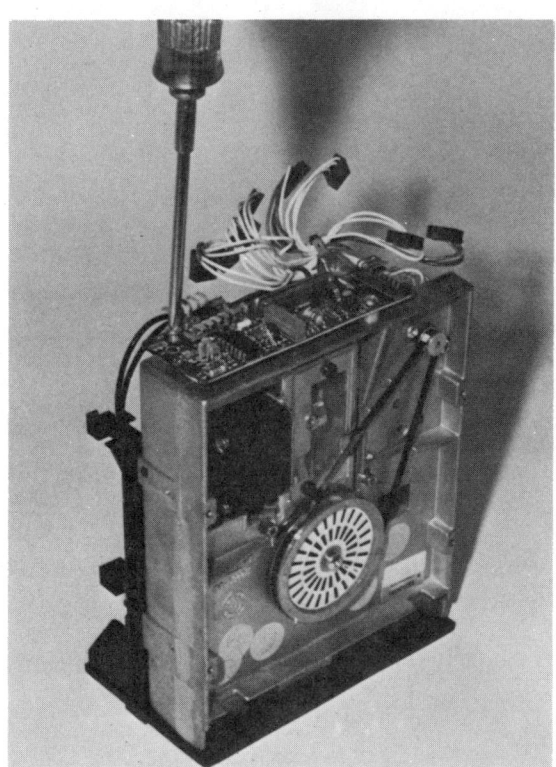

FIG. 4–39   Step 5: On Tandon
drives, unscrew the servo board
from the back of the drive.

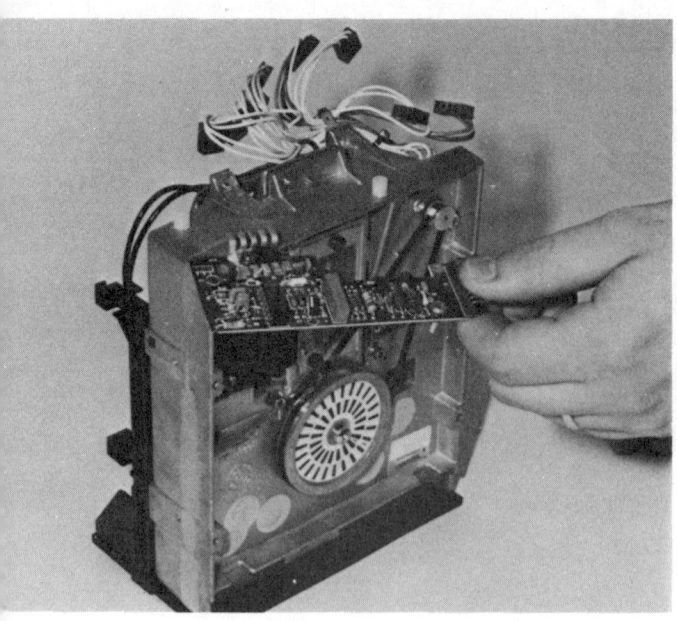

FIG. 4–40   Carefully remove the servo board and set aside.

==

Troubleshooting the Boards
5

There are many things on the circuit boards that cannot be tested without some very expensive equipment. However, you can easily test one very critical circuit function—memory. Happily this is the most common failure. It's also one of the more critical failures, since a failing memory chip can destroy hours of work.

Each time you turn on the computer, it goes through the self test. After checking the power supply, the self diagnostics routine goes inside the system board and then into memory. The diagnostics diskette does the same thing, and in the same order. In both cases, the more memory you have in the computer, the longer the testing will take.

As mentioned in Chapter 2, there are devices that defeat the self test. The idea behind these devices is to allow the user to perform a hardware reset without going through the whole test again. (If the "Ctrl-Alt-Del" reset is disabled by the software, the system will lock up and then the user has no choice but to shut down the power and begin all over again.) Many people use the self test bypass device to avoid the minute or so required to test a fully loaded memory. Again, this is not a proper solution. If the memory is malfunctioning, you want to know it *before* you begin to work, not after it's too late.

SYSTEM BOARD TYPES

You've probably heard the system board called the "mother board." Technically, this term isn't appropriate because "mother board" is used by Apple to describe their own system board.

FIG. 5–1 The IBM PC system board. The label identifying the capacity of the board is in the middle of the far left side.

The PC has two kinds of system boards, depending on when the computer was built. Earlier models can hold up to 64K of RAM on the main board. If you want more memory you have to use expansion cards. In more recent models, the system board can hold up to 256K of RAM. The upper limit on the older system board is 544K. The newer system boards can be expanded up to 640K.

Both types of system boards work well. The major disadvantage of the earlier board is that you have to use one of the five expansion slots for the added memory. (This upgrading also costs a little more than it does on the newer boards since an expansion card must be purchased in addition to the RAM chips.)

The next time you remove the cabinet of your computer, take a look at the left-hand side of the system board. Along the edge near the middle will be a label identifying the capacity of the board. If it says "16KB–64KB CPU" you have one of the earlier boards. The newer board shows "64KB–256KB CPU." (If you've already added memory to your computer, you'll probably know which one you have.)

TESTING THE SYSTEM BOARD

During the self test, the computer will automatically check the system board. An audio signal of one long beep and one short beep indicates a problem with the system board. The error code for the system board is 1xx (with the x's being any numbers). A reading of 100 (which appears only when you use the diagnostics diskette) means that the system board has tested successfully.

If you get an error reading of 199 while running diagnostics, don't worry. This simply means that you pressed "N" when asked if the devices installed are correct. An error other than this means that something serious could be wrong.

Unfortunately, if something *is* wrong with the system board, you won't be able to do much about it other than take it to an experienced technician who has the right test equipment.

Check the system board carefully to be sure that all connectors are secure. *Shut off the power first!* Unplug them and reinsert, then run the test again.

If things still don't function properly, get out your meter and set it to read 12 volts dc. You'll be testing the power flowing into the system board by measuring across various pins on the strip of 12 pins that go from the power supply to the system board. This strip is located toward the rear of the computer and next to the power supply. Pin 1 is farther back. Pin 2 is empty.

With the power on, touch the common probe (black) to pin 5 and the vom probe (red) to pin 1. A reading of anything from 2.4 to 5.2 volts is okay. The

FIG. 5–2 Location of system board power connector.

power supply could be bad if the reading you get is outside this range (see Chapter 6).

If you get an acceptable reading, check the voltage across the other pins as in Table 5-1. If these match those given in the table, the power supply is providing the correct voltages. This means that the system board is at fault. If the readings *don't* match those in the table, the power supply is probably at fault. Go to the Chapter 6 section on "Power Supply" and run that series of tests.

If the voltage reading you get between pins 1 and 5 is *incorrect*, remove all options and adapter cards from the system board, unplug the power connector, and set your meter to read ohms on the x1 scale. With the power off, check the resistance across the pins as given in Table 5-2. Keep in mind that you're looking for minimum ohms. The readings you get *can* be higher. This test measures the resistance of the circuitry of the system board and tells you if a component has shorted so as to give a resistance reading that is too low.

If you get a resistance reading *lower* than given in the Table 5-2, make a note of what the reading was and where it occurred. (To be safe, test the spot again.) This means that a component (or components) has failed, and that the system board is faulty. *Correct* readings all the way across mean that the power supply is suspect. (Go to Chapter 6.)

CHECKING THE OBVIOUS

Before you go to a lot of trouble testing the system board, however, make sure you've eliminated other possible causes of trouble. Open the cabinet and visually inspect the inside. Is there anything that could be causing an accidental

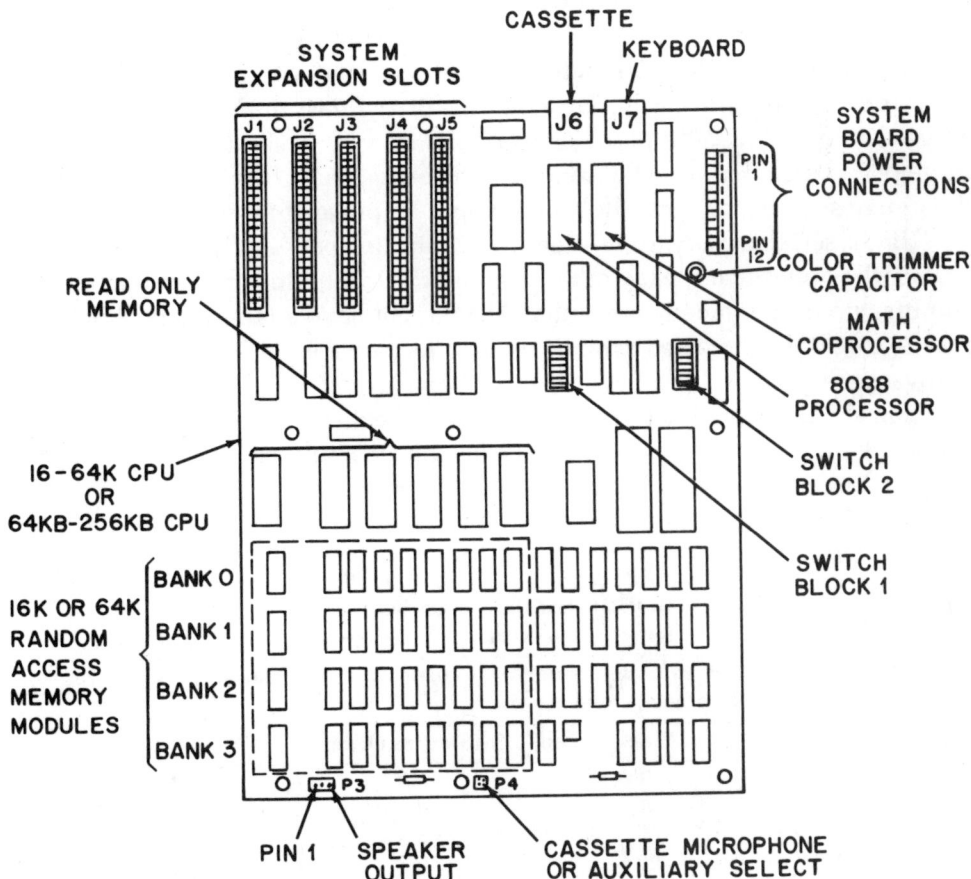

FIG. 5–3 Pin locations on system board power connector.

Williams: Repair & Maintenance of Your IBM PC (Chilton)

TABLE 5-1
Voltages at the System Board
Power Connectors

Black	Red	Voltage (dc)
5	1	2.4–5.2
5	10	4.8–5.2
4	8	10.8–12.9
7	3	11.5–12.6
9	6	4.5–5.4

short (like a screw that has fallen onto the board)? Are all option and adapter cards plugged firmly into place? Are all the cables and connectors secure? (For both things, unplug and reinsert with the power off.)

Outside the cabinet, are the plugs and cables secure? Even more important, are they in the right places? Inside the computer you can't plug something into the wrong place very easily. Outside it's all too easy to have the keyboard plugged into the cassette port, for example. (If your computer does not have these ports labeled, the keyboard port is the one closer to the power supply and the power switch.) Or a parallel cable might be plugged into a serial port.

IBM's serial port is a male end; the parallel port is a female end. It is exactly the opposite on many other computers. Add to this the fact that options manufactured by someone other than IBM might have a female serial port. It's easy to forget this reversed scheme and try to plug a serial device into what is really a parallel port.

You should know the purpose of each expansion card. For example, a circuit card that is meant to allow you asynch communications (as through a

TABLE 5-2
Resistance at the System Board
Power Connectors

Note: Shut off power and remove connector before making measurements. Set your meter to read in the x1 ohms range.

Black	Red	Minimum Ohms
8	10	0.8
8	11	0.8
8	12	0.8
5	3	6
6	4	48
7	9	17

Williams: Repair & Maintenance of Your IBM PC (Chilton)

TROUBLESHOOTING THE BOARDS

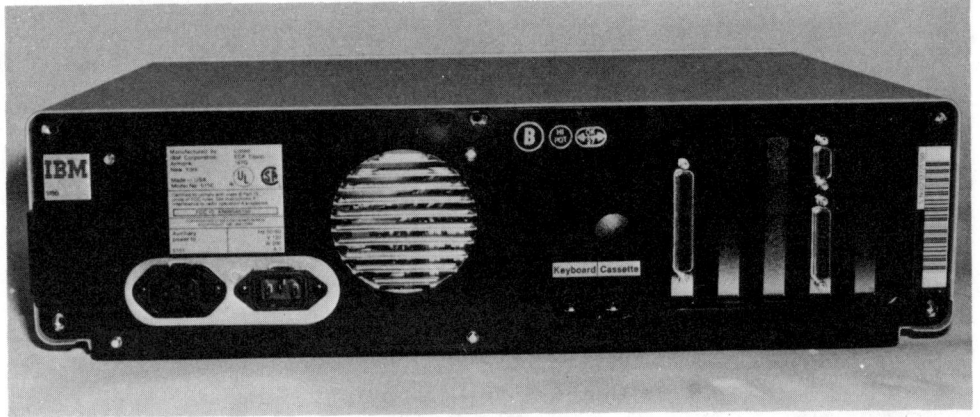

FIG. 5–4 The back of the PC. Be sure that everything is connected to the correct ports.

modem) has a serial port. Since you know that the port is serial, you won't have to worry about recognizing it as being female or male (except when it comes time to connect it to the external device).

If everything seems fine, disconnect all external devices and try the self test again. If the error disppears, it was caused by one of these devices. Plug them in one at a time (with the power shut down *each time.*) If the error does not disappear, remove all internal options and run the self test.

SWITCH SETTINGS

Unless you've made an addition to your system, the switch settings are probably just fine. It's unlikely that you've bumped them out of position, and even less likely that one of the two switches has failed.

These switches are located approximately in the center of the system board and near the power supply (Figure 5–5; see also Figure 5–3). The switch closest to the power supply is switch number 2. They are also labeled on the circuit board if you look closely.

Making additions usually requires a change in the switches. If you don't reset the switches correctly, the result can be anything from the computer refusing to accept the addition to some very strange malfunctions and error codes.

Additions that require a change in the switches generally have documentation that shows you what the changes are. The *Guide to Operations* manual also shows the switch settings. To make things even easier, correct switch settings for various options are located at the end of this book.

FIG. 5–5 System board switch locations.

If you've never looked at the switches before, take a moment to do so. Knowing what the settings are *supposed* to be when everything is operating correctly can help if you ever make a change in the system or if something ever goes wrong.

Note: If the self test and diagnostics indicate that the added RAM memory is functioning and is of the correct amount, and one or more programs fail to work (ones that have worked before), check the switch settings both on the system board and on the expansion card. This is especially important if you have more than one memory expansion card.

MEMORY

RAM means random access memory. RAM allows you to throw data into the memory and to retrieve them in any order. Without this your computer would be able to do very little. RAM is measured in bytes, or more often in kilobytes ("K" for thousands of bytes). Generally, the more RAM you have, the more you can do.

The newer IBM PCs can directly address (get at) 640K. (The older units can access only 544K of RAM.) The CPU of the PC is capable of addressing more than this, but the PC is set up so that at 640K it begins to look for the built-in ROM (Read Only Memory). If you've added more than 640K of RAM, the computer will simply ignore the extra RAM and go for any available ROM. The switches on the system board will only allow you to tell the computer that it has 640K of RAM to prevent such a complication.

The memory board is set in rows, called *banks*. One bank equals 64K of RAM, with nine chips making up each bank. (The older 16KB–64KB system board is different. See below.) Eight chips are used for the data. Each of these represents a particular data bit. They are numbered from 0 to 7. The module labeled "P" is the parity, or error checking, bit. On the system board this module is set slightly apart from the row of data bit modules. Knowing the location of each bank and bit is necessary if you're going to track down a failing memory module.

The RAM on the system board is located toward the front and on the left (see Figures 5–6 and 5–3). The RAM modules are placed in four rows of nine modules each. In each row, the module farthest to the left outside is the parity bit module. The other eight are data bit modules.

FIG. 5–6 Location of RAM on system board.

The difference between the older 16KB–64KB system board and the newer 64KB–256KB system board is in the modules used. With the older boards, 16K chips are used. (You cannot simply change the chips to upgrade the system board. There are some other changes required to make the newer boards work.) In appearance, the two boards are identical. The diagnostic codes are slightly different, since the testing of the older board must also tell you which row to check as well as the bank. (More information on this is contained under "Diagnostics" below.)

If you don't already know how your memory board is set up (on both the system board and on any memory expansion cards), remove the cabinet of your PC and take a look. Most boards have the banks and the bits clearly labeled. The only trick is knowing that when an expansion card has "Bank 0" printed on it (or labeled as such in the documentation), it really isn't. Bank 0 is always on the system board.

You can have up to nine banks of memory in the computer system. The first bank is given the number 0, with the subsequent banks being numbered 1, 2, and so on. As already mentioned, the earlier system board will have just bank 0 on it. Expansion cards with memory will then begin with bank 1. The newer system boards can hold 256K on the system board itself, which means that it can hold four banks of RAM. These are numbered from 0 to 3, with bank 4 being the first on an expansion card. Again, be aware of this. The expansion card (or its documentation) will list its own banks with numbers. These may not be the bank numbers that the computer sees.

The first banks are always on the system board. If you have the older 16KB–64KB system board, bank 0 is on the system board, with bank 1 being the first bank on an expansion board. If you have the newer 64KB–256KB system board, banks 0 through 3 will be on that system board, with bank 4 being the first on an expansion card.

Note: Before adding a memory expansion card, the memory on the system board must be full. The same rule applies to adding a second memory board. Before adding the second board, the first board must be full.

DIAGNOSTICS

During POST and diagnostics, the display should show a failing module with a code number. (A 200 during diagnostics means that everything is fine.) Normally a four-digit code will be given, followed by the number 201. Those four digits will tell you exactly where the problem is. (The 201 indicates a memory malfunction. A 201 error code by itself is often an indication that the switches are not set correctly.)

TABLE 5-3
RAM Module Error Codes

Note: The first digit(s) of the error code indicates which bank the failed module is in (see text). The third and fourth digits indicate which module is bad. This table applies th both the 64K and 256K system boards.

3rd and 4th Digits	Bit Module	3rd and 4th Digits	Bit Module
00	P	10	4
01	0	20	5
02	1	40	6
04	2	80	7
08	3		

The first digit of the four-digit code tells you which bank the module is in. For example, if the error code begins with the number 3 (as in "3020 201") the failure is in bank 3. The last two digits tell you which bit module is bad (see Table 5-3). Using the above example code, the 20 indicates that module 5 is malfunctioning.

It's as simple as that. If you're not sure how this works, try a few yourself. Write the "address" of several modules using Table 5-3 as a guide.

If you have the older 16KB-64KB system board, the memory installed is in 16K chips instead of 64K chips, and the error code used to locate the failing module begins a little differently.

The first digit of the error code will be 0 if the module at fault is on the

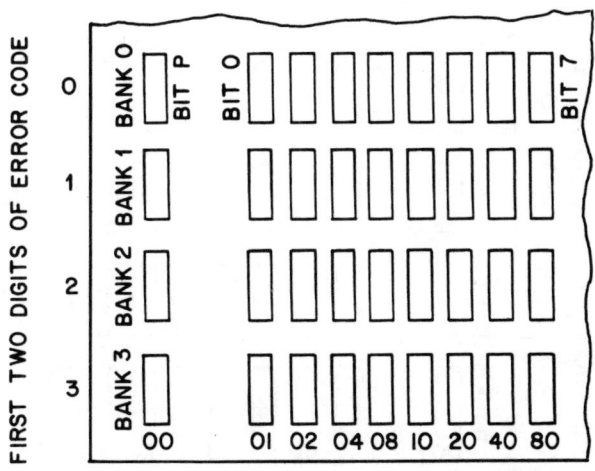

FIG. 5-7 System board RAM module addresses, newer board (64KB-256KB).

system board. Although it appears that the system board has four banks, remember that they are *all* bank 0 on the 16KB–64KB board, with each row making up a fourth of the total 64K (4 × 16K = 64K).

The row farthest back is named 00. Next comes 04, then 08, and finally 0C. Thus an error code of 0020 201 indicates that the fault is on the system board, in the first row, and the fifth bit module. (See Figure 5–8.) The code 0004 indicates bank 0 (on system board) and data bit module 2. The code 0880 indicates bank 2 and data bit module 7.

Note: If the error code does not correspond with any address in Table 5–3, first check the switch settings. If this doesn't correct the problem, it is likely that more than one RAM module is bad.

Memory expansion cards are diagnosed exactly the same way (see Figure 5–9). The error code will indicate a bank that is not on the system board (bank 5, for example) and then the bit module that is faulty.

Before anything else, check all switch settings. There are the two DIP switches on the system board that must be set in order for the computer to "see" the added memory. There are also DIP switches and other settings on the expansion board. Refer to the documentation with your expansion board for the correct settings on the board itself.

TESTING THE MODULES

When you've located the failing module, you can test it by swapping it with another. The easiest way to do this is to switch it with another module in the

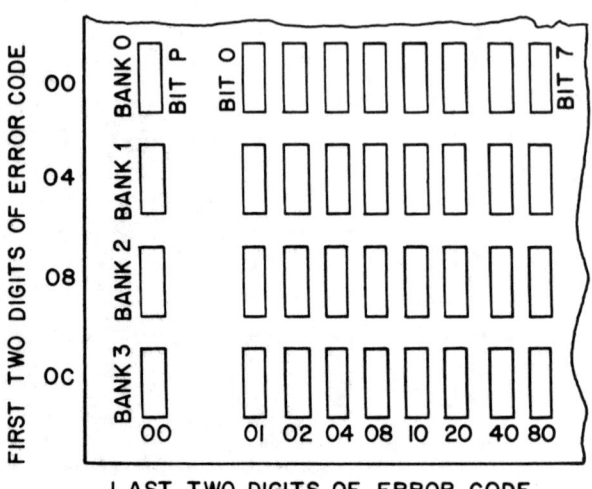

FIG. 5–8 System board RAM module addresses, older board (16KB–64KB).

Williams: Repair & Maintenance of Your IBM PC (Chilton)

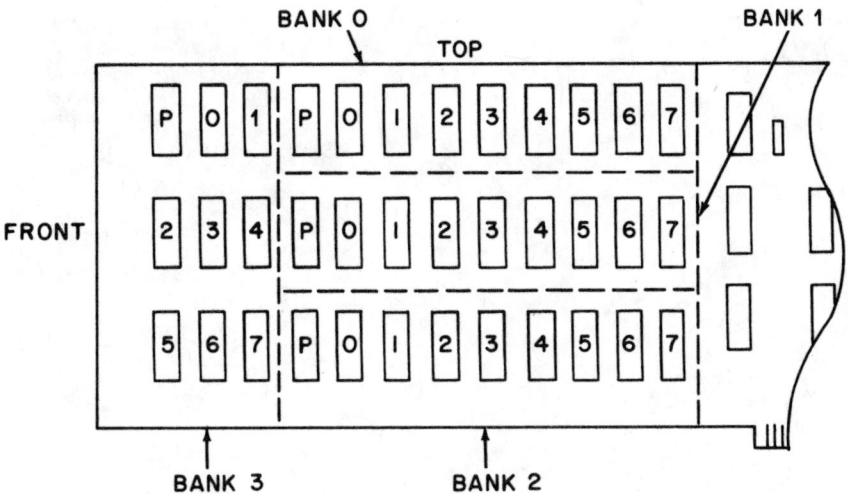

FIG. 5—9 Expansion RAM card module addresses. Note: Your board may be different. Check the documentation that came with the board.

same bank (Figure 5-10). When you run the test again, the error code should show that the fault has changed positions. For example, if diagnostics indicates that module 5 has failed, swap it with module 1 in the same bank. The new error code should now read 3002 201. If it doesn't, try another swap within that bank. For example, swap the suspected module now in slot 1 with the module in slot 2. The error code should now read 3004 201.

If this doesn't change things, the bank might have a more severe failure. Try a swap with another bank that seems to test okay.

Diagnostics tests the banks one by one, starting with the lowest bank. If the test shows an error code with a low-numbered bank, *it will not have tested the banks above.* If the failure has occurred in one of the higher banks (or on an expansion card), you can eliminate the possibility of a system board failure by simply switching out those higher memory banks. (To do this, reset the two DIP switches on the system board according to the settings given in the *Guide to Operations* manual or at the end of this book.)

OTHER BOARDS

Without sophisticated test equipment, you probably won't be able to find the specific problem with a circuit board. However, you can perform a few simple tests on your own.

Make a note of the error code and/or what is happening (or not happening). This should isolate the problem to a single board.

FIG. 5–10 The easiest way to test a RAM module is to switch it with another. In extreme cases, switch all modules in a bank with those in another bank.

The switch settings on the system board are designed for IBM devices. It is possible that a non-IBM device will show an error even if everything works normally, the board is in perfect condition, and the switches are set correctly. Check the manual or other documentation that came with the device for correct installation and switch settings. If you have a non-IBM board, an error indicated by diagnostics could be an error in itself. You can check a non-IBM board by taking it to the dealer you bought it from for testing.

Obviously, if the board has been functioning perfectly for some time without an error code showing, you won't suspect a mismatch of the manufacturer to the diagnostics.

First visually check the suspected board. Is it snugly connected? Are all cables going to it tight? Have any of the components worked loose? *(Shut down the power before removing or inserting any board or device.)*

Unplug the board and run the diagnostics again. If the error code disappears, you'll know that the board is at fault. If the error is still there, the problem is elsewhere in the system. You can also disconnect all external devices attached to the board. If the error disappears with the card in place but with it disconnected with everything else, the problem *could* be in that external device.

Williams: Repair & Maintenance of Your IBM PC (Chilton)

TROUBLESHOOTING THE BOARDS

Try cleaning the contacts with a high-quality electronic cleaner (*not* a television tuner cleaner—don't use anything that could leave a residue!). If you don't have a cleaner around, you can use a soft pencil eraser, making sure that you don't get any particles inside the computer (Figure 5–11).

SUMMARY

Diagnosis of board malfunctions begins by determining which function is acting up. If a drive is doing strange things, don't waste time testing the RAM board. If the printer is acting up, don't waste time fiddling with the driver adapter.

By observation and note taking, you should be able to isolate the malfunctioning board quickly.

As always, take some time to visually inspect for the obvious. Are the

FIG. 5–11 Cleaning the contacts.

switch settings correct? Are all cables and wires firmly attached? Is the board compatible with the PC? Has it changed something else (e.g., have you made an addition that gives you more serial or parallel ports than can be supported by the computer)?

Most of the time repair is done by replacement. This is because tracking the failure to a single component usually requires a lot of time and some special equipment.

RAM boards and the memory modules on them can be tested quickly and easily. The computer will do most of the job for you by finding the module (or modules) at fault. From there it is simply a matter of swapping the modules with good ones.

Fortunately, if a board operates when you first install it, it will probably continue to function for many years. No maintenance is required other than to change the battery for a built-in clock.

```
================================
```

Power Supplies, Keyboards, Printers, and Monitors
6

You can connect a number of different options and devices to your PC. Since there are so many different options, so many different manufacturers, and so many variables, it is impossible to cover them all. We have included only the most common options in this chapter. These should give you some basic guidelines for repair and installation.

The diagnostics diskette will help to isolate problems in many of the options available. Use this and the information in Chapter 2 before you begin poking around in the machinery. The steps taken in Chapter 2 are what should have brought you to this chapter. (Chapter 2 might reveal that the problem is in a place you didn't suspect at first.)

As mentioned in Chapter 2, some options manufactured by companies other than those "approved" by IBM will give false readings during diagnostics. Keep this in mind while trying to find the problem. For example, the diagnostics might indicate that there is a problem with the keyboard, when in fact there is no problem at all in that device.

THE POWER SUPPLY

The driving force behind the PC (and all things electronic) is the power supply. It does just what the name implies. It takes the 120 volts at 60 cycles per second from the wall outlet and changes it to a clean, steady 5 and 12 volt dc supply.

Normally it does its job just fine and gives no problems. Unlike the power

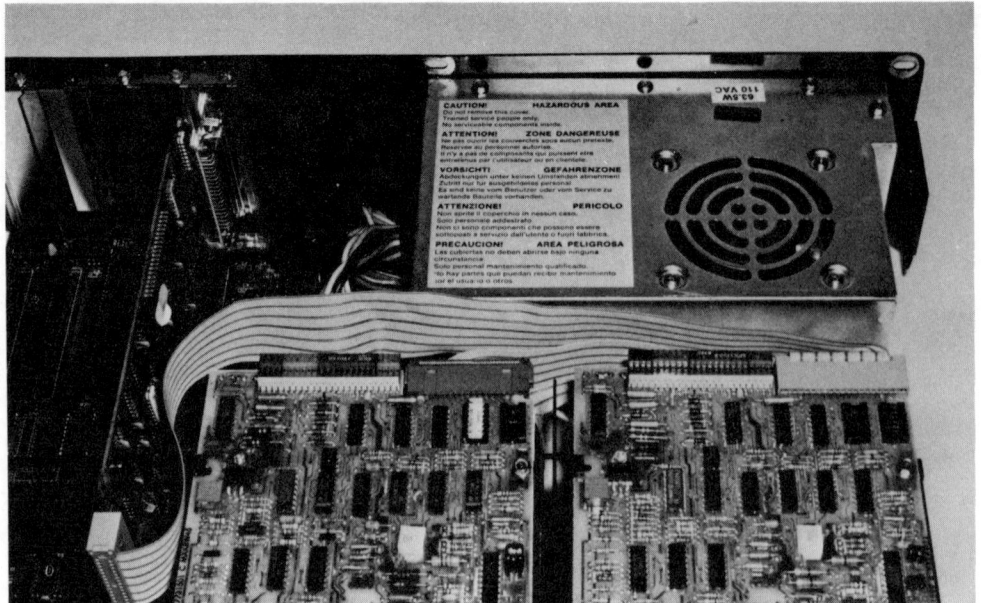

FIG. 6-1 Location of the power supply.

supplies of some other computers, the power supply of the PC is relatively tough. Chances are you'll never have to worry about it.

Note: The fusing is handled by a Type 2 SOC SD4 fuse that automatically trips when the power draw goes beyond the limits. It then resets itself automatically when the power supply is shut off for 5 seconds. With some power supplies you cannot get at this fuse, and normally you should not try. Do not open the power supply unless you know what you're doing.

If the power supply *is* acting up, you may not be readily able to spot it as the source of the problem. The computer might seem to be completely dead. This could be caused by the power supply, but it could also be a problem with the system board or from a combination of other things. Your PC could also seem to be operating normally except for a data read/write problem. You might be inclined to automatically place the blame on the memory board or on the drives, while the actual cause *could* be with the power supply. Don't replace anything until you know for sure.

POWER SUPPLY DIAGNOSTICS

If nothing happens when you flip the power switch, you are likely to accuse the power supply without further thinking. This symptom could mean that the power supply has died. It could also mean that something else is wrong.

FIG. 6–2 The power supply.

Remember that the fan is wired directly to the incoming 120-volt line. If the fan isn't working, chances are good that the problem is outside the computer. (It would be rare to have both the fan and the power supply give out at the same time.) If the fan is operating, power is getting to the computer.

The next time you switch on your computer, listen carefully for a slight click coming from the speaker. This click means that power is getting to the

TABLE 6–1
Power Supply Specifications

Input Power	104–127 volts ac (120 nom.) @60 Hz; 2.5 amps
AC Output	101–130 volts ac (120 nom.) @60 Hz; 0.75 amps
DC Output	+5 volts dc; 2.3–7.0 amps
	−5 volts dc; 0–0.3 amps
	+12 volts dc; 0–2.0 amps
	−12 volts dc; 0–0.25 amps

Williams: Repair & Maintenance of Your IBM PC (Chilton)

POWER SUPPLIES, KEYBOARDS, PRINTERS, AND MONITORS

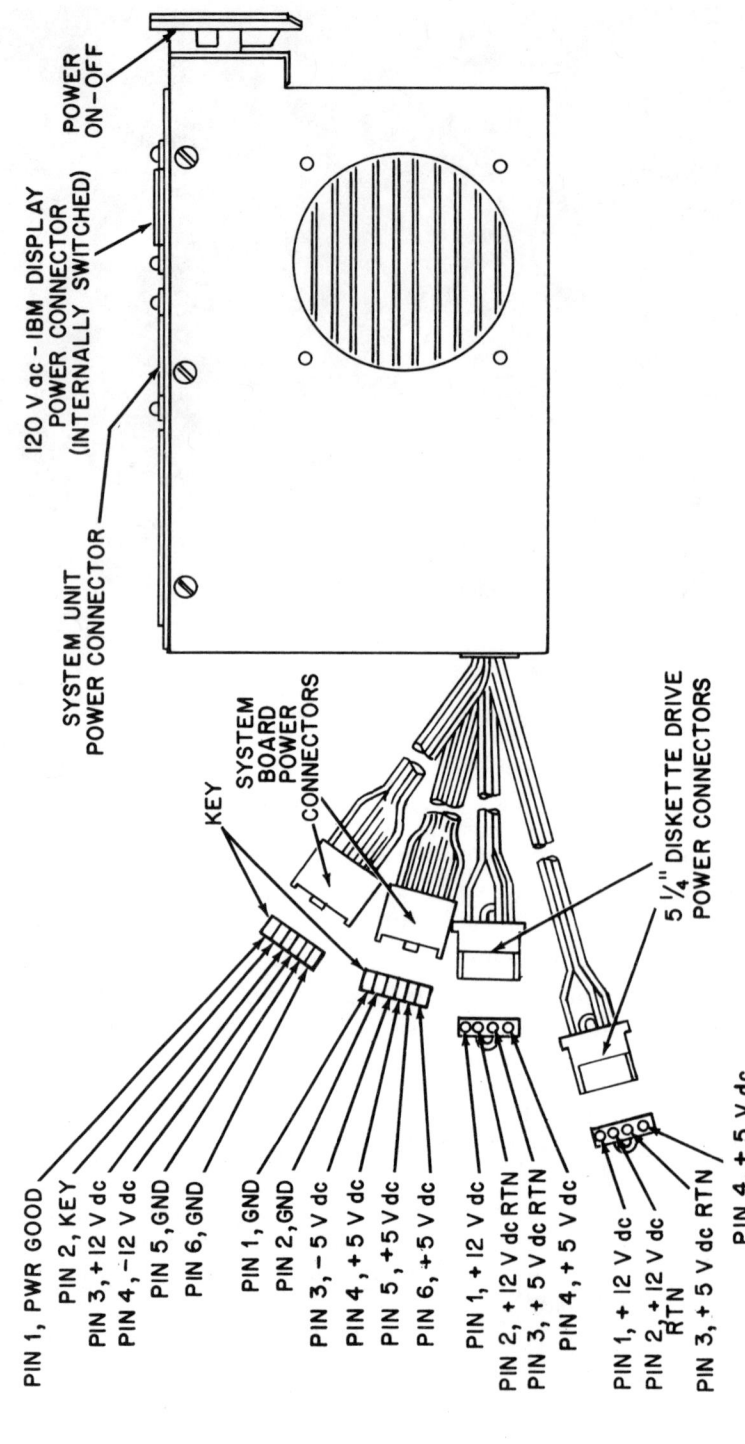

POWER ON-OFF

120 V ac - IBM DISPLAY POWER CONNECTOR (INTERNALLY SWITCHED)

SYSTEM UNIT POWER CONNECTOR

KEY

SYSTEM BOARD POWER CONNECTORS

5 1/4" DISKETTE DRIVE POWER CONNECTORS

PIN 1, PWR GOOD
PIN 2, KEY
PIN 3, + 12 V dc
PIN 4, - 12 V dc
PIN 5, GND
PIN 6, GND

PIN 1, GND
PIN 2, GND
PIN 3, - 5 V dc
PIN 4, + 5 V dc
PIN 5, + 5 V dc
PIN 6, + 5 V dc

PIN 1, + 12 V dc
PIN 2, + 12 V dc RTN
PIN 3, + 5 V dc RTN
PIN 4, + 5 V dc

PIN 1, + 12 V dc
PIN 2, + 12 V dc RTN
PIN 3, + 5 V dc RTN
PIN 4, + 5 V dc

FIG. 6-3 Power supply connectors and outputs.

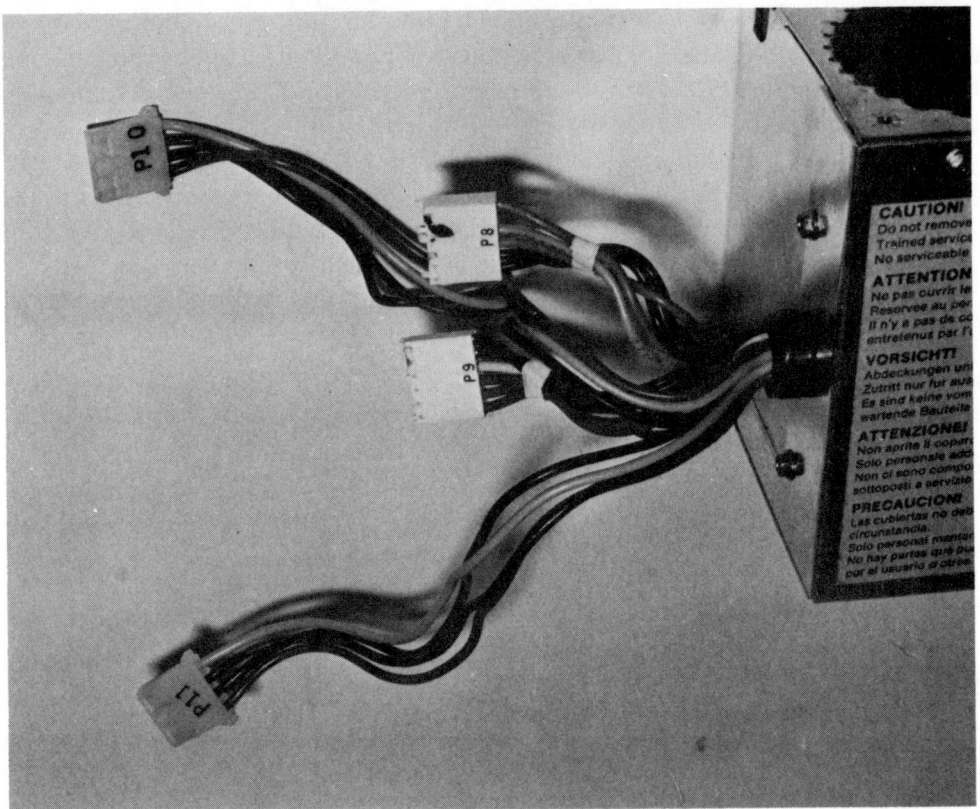

FIG. 6–3 *continued.*

system board. If you hear the click, chances are good that the power supply is functioning properly and that the problem is elsewhere in the system. (This isn't *necessarily* true—just a general guideline.)

1. CHECK INCOMING POWER

If power is obviously present (i.e., the power-on self test operates and/or the fan is running), you can skip this first check and go to Step 2. If nothing at all is happening, begin with this step.

The first thing is to check the plugs, the outlets, and the power switches. *(Shut off the power before checking the connectors. Otherwise you can cause considerable damage.)* Once you have done this and know that power is getting to the power supply, you have eliminated the obvious. You've also started the process of isolating the problem.

As mentioned in Chapter 2, you can use a lamp or any number of other things to check the outlet for power. You are better off checking it with a meter. The power supply operates in the range of 104 volts and 127 volts. A lamp will operate beyond these ranges without any apparent difference, but the power supply of the PC will automatically shut itself down (everything except the fan). Until you've checked the outlet with a meter, you won't know for certain if the problem is in the computer or in the lines that supply the outlet. (More information on using a meter to check outlets is contained in Chapters 1 and 2.)

If power is getting to the power supply but nothing is happening move on to Step Four.

2. SELF TEST AND DIAGNOSTICS DISKETTE

A POST error code of "System Unit 02x" (with the "x" being any number), a continous beep, or a series of short beeps indicate a *probable* problem with the power supply. These signals will occur during the power-on self test and again while running the diagnostics diskette (if you have power to run diagnostics, of course).

This step is critical if there is power and you can perform the step at all. If you have to skip this step temporarily, return to it as soon as power has been reestablished. It will show that the power supply isn't functioning properly (identified by the above signals). It will also help indicate if there is an additional problem elsewhere in the system. Due to its built-in protective circuitry, the power supply will shut down if something else in the computer is faulty.

3. CHECK EXTERNAL DEVICES

As just mentioned, the power supply in the PC has the ability to shut itself down if the power drawn from it is the incorrect value. This protects both the power supply and the devices. It can also lead you to believe that the power supply is dead when in fact it is doing just what it was designed to do.

If the problem is in one of the connected devices (printer, keyboard, external modem, etc.), you can find out simply by disconnecting these and checking for power again. Shut down all power and disconnect everything that is connected to the computer. Apply power again. If there is still no power, go to Step 6. If you have power flowing again, you know that one of the devices is at fault. To find out which one it is, merely plug them in one at a time (with the power shut down each time) until the power fails again.

Shutting down the power before connecting or disconnecting *anything* is

Williams: Repair & Maintenance of Your IBM PC (Chilton)

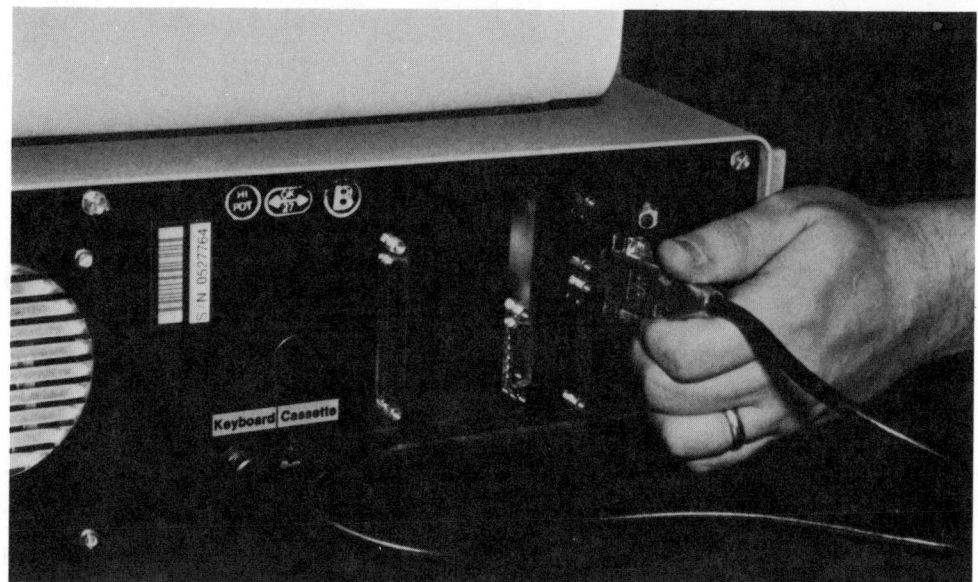

FIG. 6—4 Disconnect all external devices, including the keyboard, and try to power up again.

extremely important. If you don't shut down the power you could end up with a destroyed computer.

4. CHECK INTERNAL DEVICES

If disconnecting external devices doesn't help, the next step is to apply the same idea to internal options. With the power off, disconnect all internal options, including the drives and the option boards. If power now flows, you know that one of these is causing the trouble. Shut off the power and reconnect these devices and options one at a time until the trouble is found. Again, remember to shut off the power *each time* before making a new connection. You don't want to destroy your computer for the sake of the second it takes to shut down.

Once you know that all other internal devices are functioning properly (as far as the power supply is concerned) test the disk drives. Since the two drives are connected together, they are tested one at a time. Disconnect drive B to test drive A. If things seem to be okay, disconnect A and reconnect drive B. (A short in the cable can be detected by checking drive A and then drive B on each of the two cables. Both power and signal cables can be tested by this swap method.)

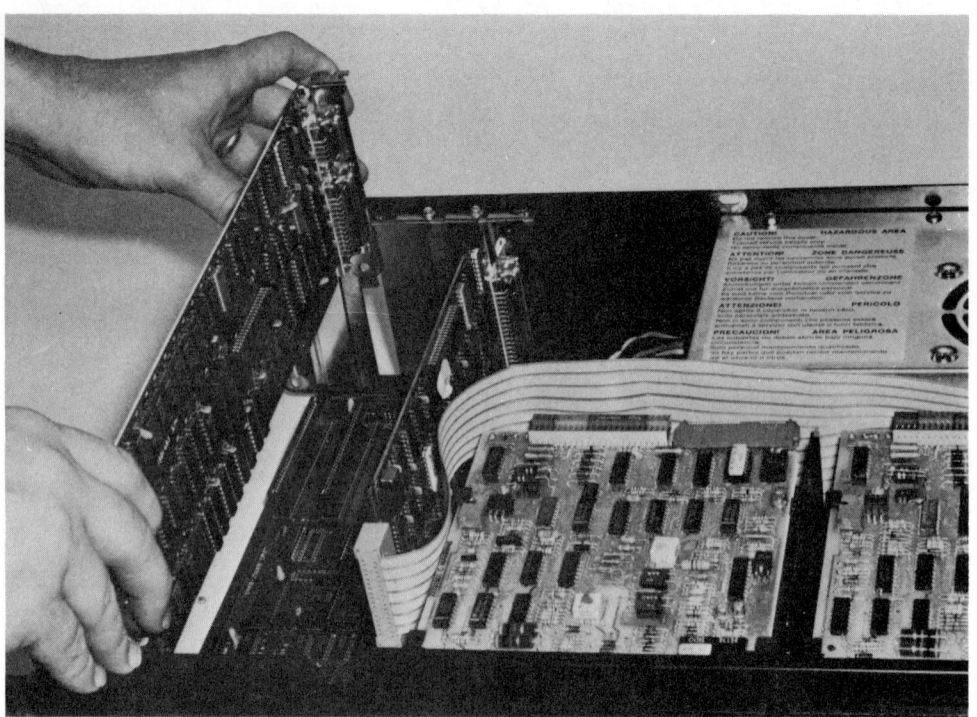

FIG. 6–5 Disconnect all internal options, *with the power off!*

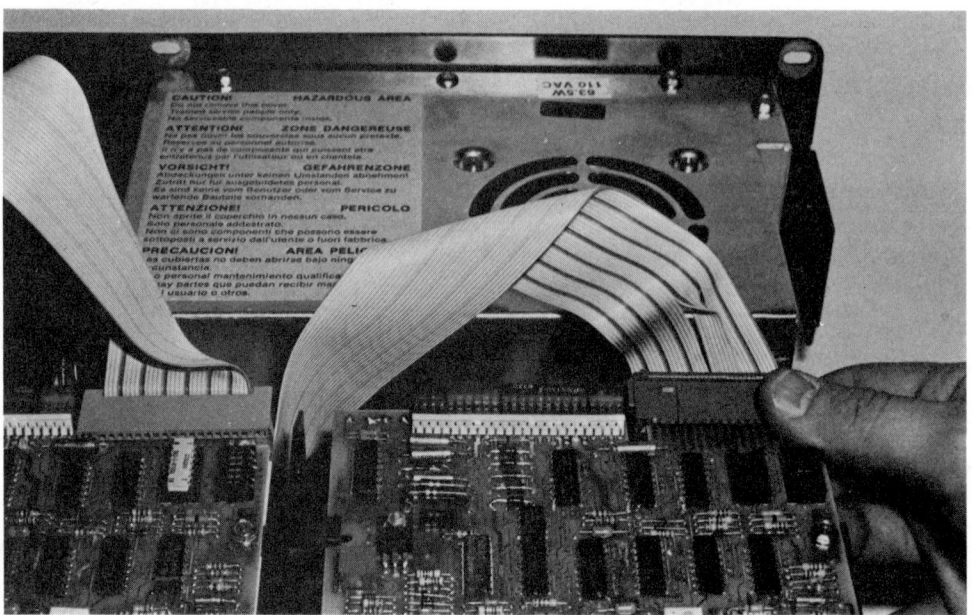

FIG. 6–6 Disconnect drive B to test drive A. Disconnect drive A to test drive B.

Williams: Repair & Maintenance of Your IBM PC (Chilton)

POWER SUPPLIES, KEYBOARDS, PRINTERS, AND MONITORS

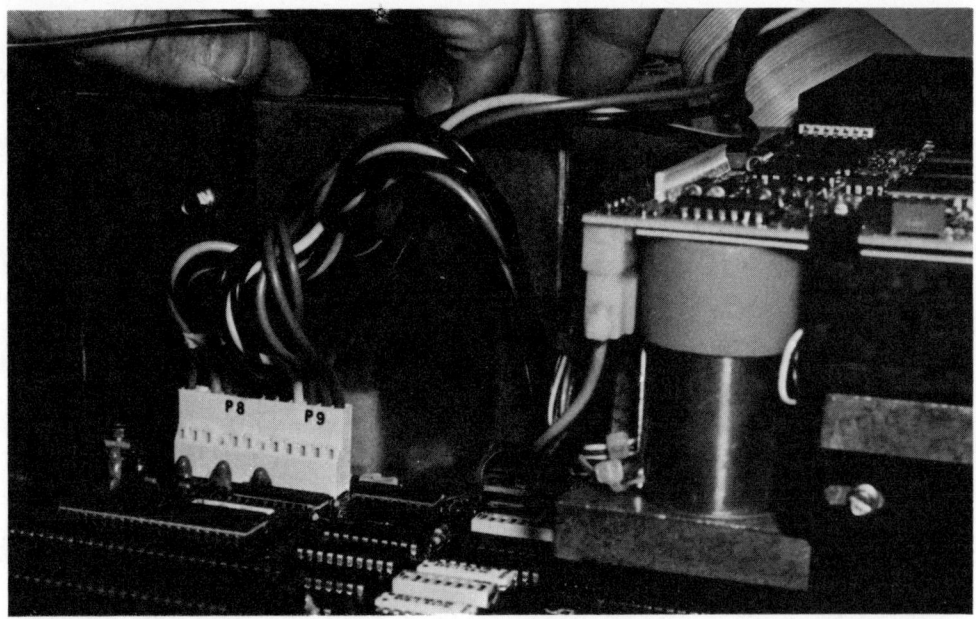

FIG. 6–7 System board power connector. Voltage between pin 1 and pin 5 (common) should be between 2.4 and 5.2 volts dc. (See Figure 6–8.)

If a device is suspect, go to the appropriate section in this book for further help (e.g., if the offending device is the memory board go to Chapter 5, "Troubleshooting the Boards").

If nothing changes after removing all internal devices and options from the power, go to Step 5.

5. CHECK VOLTAGES TO SYSTEM BOARD

This step involves the use of your multimeter set to the 12 volt dc range. The test points are where the cable from the power supply plugs into the system board (see Figures 6–7 and 6–8). Leave the system board connected. *Since the power is on during these tests, you must be extremely careful not be cause any short circuits.*

First you'll be checking the voltage between pins 1 and 5 (with pin 5 being "common"). You should get a reading of between 2.4 and 5.2 volts. If you *don't* get this reading, move immediately to Step 8. If you *do* get a correct reading, check the other pins. Figure 6–8 shows the pin locations. Refer to Table 6–2 for correct voltage readings on the other pins.

If these voltages aren't correct, you'll know that the problem is definitely

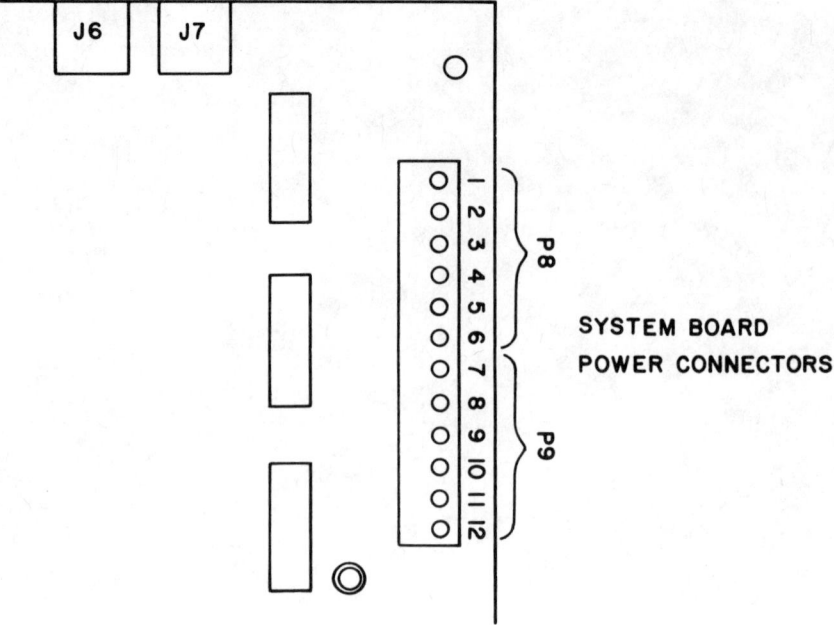

FIG. 6–8 System board power connector pin numbers: J6 is the cassette port, J7 is the keyboard port.

in the power supply. All you can do is replace the faulty unit with an exact match. If the voltages are correct, proceed to next step.

6. CHECK DISK DRIVE POWER

In Chapter 4 you learned that there are two basic types of drive: Type 1 (with a serial number that begins with A, B, or no letter) and Type 2 (with a serial number beginning with D). Testing for power is similar for both, and the pins are in the same location: top left rear when looking at the front of the drive.

TABLE 6–2
Voltage at System Board Power Connector

Black	Red	Volts dc
5	1	2.4–5.2
5	10	4.8–5.2
4	8	10.8–12.9
7	3	11.5–12.6
9	6	4.5–5.4

Williams: Repair & Maintenance of Your IBM PC (Chilton)

POWER SUPPLIES, KEYBOARDS, PRINTERS, AND MONITORS

There are four pins here (see Figure 6-9) for testing the incoming power.

The voltage between pins 2 and 4 (with 2 being common or negative) should be between 4.8 and 5.2 volts. The voltage between pins 3 and 1 (with 3 being common) should be between 11.5 and 12.6 volts. If either is incorrect, the problem lies with the power supply, which means that you'll have to replace it. If the voltage readings are correct, go either to Step 7 or Step 8. (Skip Step 7 if you had an audio response during the power-on self test or during diagnostics.)

7. CHECK SPEAKER CONTINUITY

The speaker is connected to the system board with a two-prong plug. Disconnect the plug from the system board. Set the meter to read resistance (ohms) in the x1 range. The reading across the speaker should be approximately 8 ohms. Some meters will not measure this accurately. This isn't important. You are merely checking for continuity. If the speaker shows a relatively low resistance,

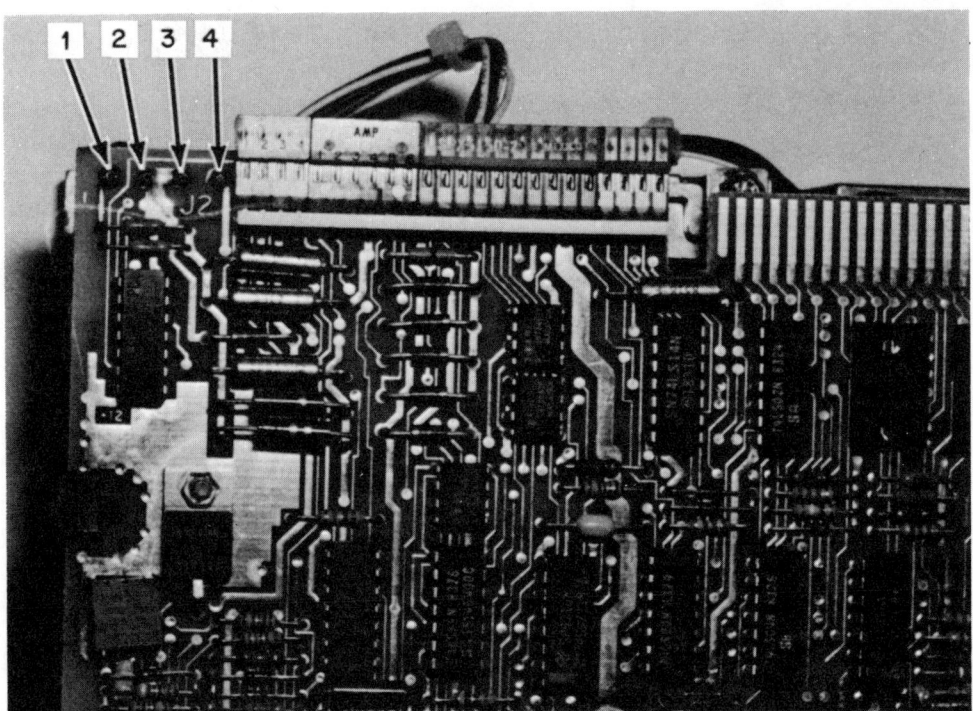

FIG. 6-9 Power connection to disk drive. Voltage between pin 2 (common) and pin 4 should be between 4.8 and 5.2 volts dc. Voltage between pin 3 (common) and pin 1 should be between 11.5 and 12.6 volts dc.

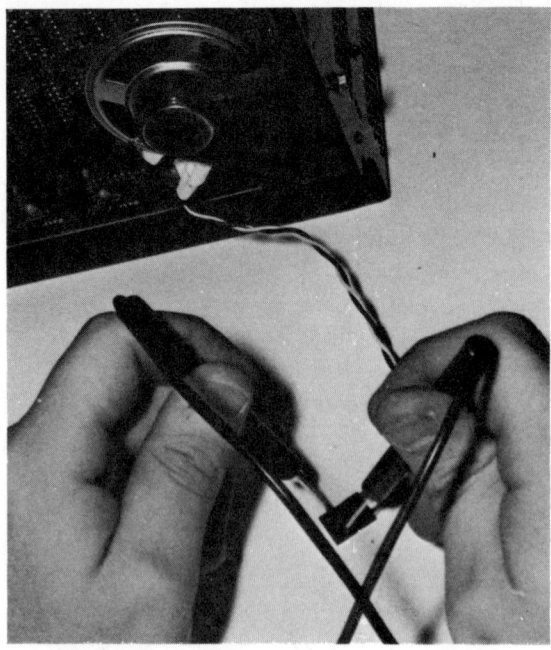

FIG. 6–10 Checking speaker continuity.

things are just fine with the speaker. If the speaker shows an infinite resistance, it means that the speaker is "open" and needs to be replaced.

Unfortunately, if the computer has passed all tests to this point, and the speaker tests okay as well, the problem is in the system board, which often means that a more expensive repair is called for. For further diagnosis go to Chapter 5.

8. CHECK RESISTANCE ON SYSTEM BOARD

If your computer has passed all previous tests you have one more thing to check: the resistance of the system board. As with all other things connected to the power supply, if the system board draws the wrong value of power it will cause the power supply to shut itself down. Some of the previous steps will have indicated this. Step 8 will verify if the problem is in the system board or in the power supply.

Remove the power connectors from the system board (Figure 6–11). This will allow you to probe the main circuits of the system board with your meter. Set your meter to the x1 range.

The pins are numbered from 1 through 12, with pin 1 being closest to the back of the computer (see Figure 6–8). The resistances in Table 6–3 are mini-

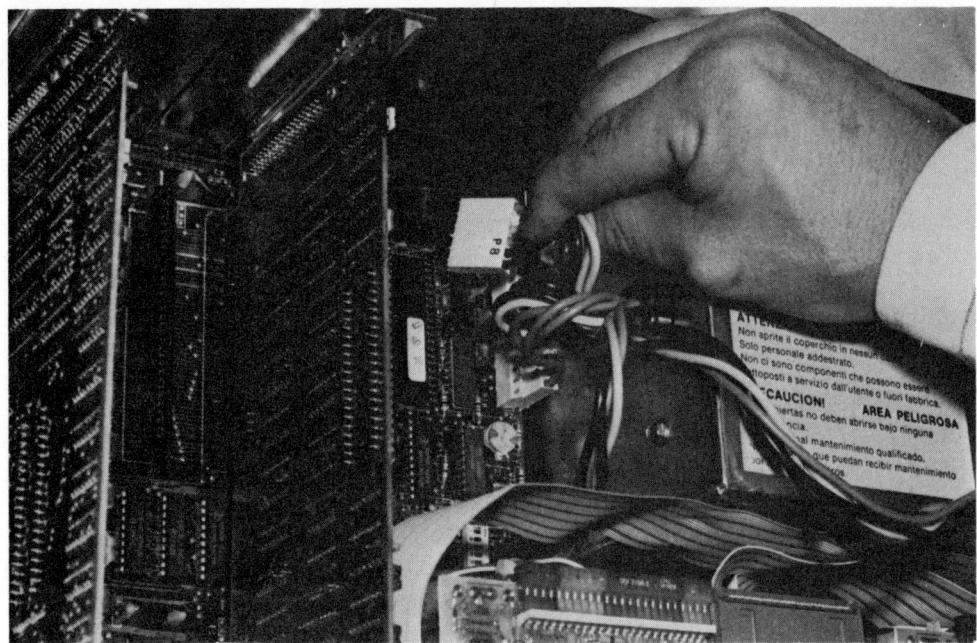

FIG. 6-11 Location of power connector to system board.

mums. The actual readings can be higher. What is important is that they are no lower than the values given in the chart.

The "com" is the black or negative lead of the meter. The "vom" probe is the red or positive lead. If the readings you get are equal to or higher than those in Table 6-3, the power supply will have to be replaced. If the readings are lower than those given, this indicates that the fault is somewhere in the system board. Go to Chapter 5 for more information on troubleshooting this main board.

TABLE 6-3
Resistance Chart of System Board

Black	Red	Minimum Ohms
8	10	0.8
8	11	0.8
8	12	0.8
5	3	6
6	4	48
7	9	17

THE KEYBOARD

There have been many things said about the keyboard of the PC; few of them have been nice. However, most of the criticisms have to do with key placement. It would be difficult to find fault with the construction of the keyboard.

Each key is a switch, but not the kind of switch you are used to. There are no contacts, as such. This means that there is no chance of arcing or other electronic wear to the keys.

Beneath each key is a small spring. The spring moves a small plastic paddle beneath the key (the little click you hear.) The motion of the paddle changes the capacitance in a cell below the key. The change in capacitance makes the keyboard work. The electronics of the keyboard read this change and send the various signals to the computer.

The spring is the only mechanism in the keyboard that wears to any extent. After a few million presses of the key, you may have to replace it. If the key clicks, the spring is probably okay. Check the sound of the click with the keys around it. If the sound is considerably different, the spring may need replacing.

The key cap comes off fairly easily (be careful not to damage it) to reveal the spring and paddle. The easiest way to get the key cap off is to make a hook from a piece of small metal wire (Figure 6–12). (The wire has to be small enough

FIG. 6–12 A wire hook is used to remove the key cap.

Williams: Repair & Maintenance of Your IBM PC (Chilton)

POWER SUPPLIES, KEYBOARDS, PRINTERS, AND MONITORS

FIG. 6–13 Key cap, paddle, and spring.

to fit between the keys.) A single hook should be sufficient, although you can make the puller as fancy as you wish.

The cap is a little tricky to reinstall, but with some patience you shouldn't have any trouble. Tip the keyboard on its side, with the wider side down. This will help the spring to go into place, and then to stay there while you snap the cap on again.

Inside the keyboard are the electronics. Those electronics handle the signal

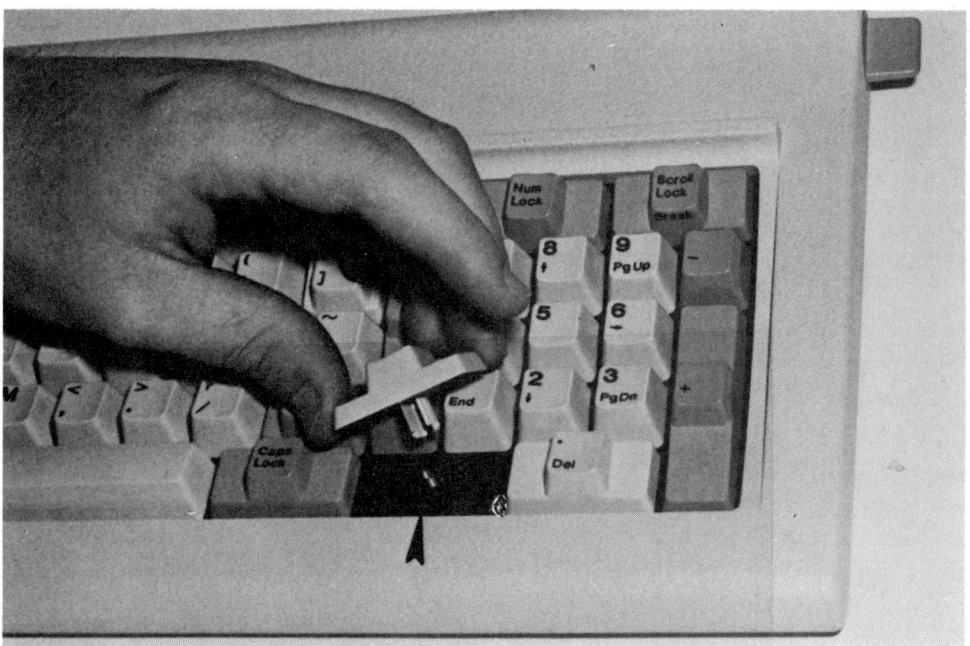

FIG. 6–14 Note location of the spring.

Williams: Repair & Maintenance of Your IBM PC (Chilton)

FIG. 6–15   Reinstalling the key cap.

from the keys and to the computer. Unlike the keys themselves, the electronics are prone to damage from fluid. To protect your keyboard, keep *all* fluids away from the board.

You've probably heard some of the horror stories about liquid falling into a computer keyboard. One of the strangest I've ever heard involves a man who decided that a cute photo would be one of his two-month old baby sitting on the keyboard in front of the computer. The child had an "accident." Obviously so did the keyboard. The acidic fluid (let's keep it clean, now) caused bad connections on the soldered components and even damaged the circuit board. This made a simple repair impossible. The whole keyboard had to be replaced.

The power-on self test runs a quick check on the keyboard each time you apply power to see if any keys are stuck down. Actually, you're testing the keyboard each time you use it as well. If one or more keys fail to operate and no error code shows on the screen, check the spring under the suspected keys.

The error code for keyboard problems is either 30x or xx30x. (Remember, the "x" represents any number.) This code should display both with the self test and with the diagnostics (although the diagnostics *does* perform a more thorough testing and might catch something that the self test didn't).

With or without this error code, if you suspect the keyboard of malfunction,

Williams: Repair & Maintenance of Your IBM PC (Chilton)

the next step is to check the voltages. To do this, remove the keyboard plug from the back of the computer. (By the way, did you check to see if the keyboard was plugged in properly in the first place? The plug fits nicely into the cassette port, which is one of the few ill-designed parts of the PC. If in doubt, remove the plug and connect it again, and while you're at it, do this same thing with all cables.)

Set your meter to read dc volts. The black (common or ground) should go to pin 4. Touching the other pins should give you a reading of between 2 and 5.5 volts (Figure 6–16). If any of these is incorrect (see Figure 6–17 for pin numbering), the problem is inside the computer, and is probably in the system board (see Chapter 5). Make a note of where the voltage was incorrect and the value you received. If the voltages are correct, the problem is either in the cable or in the keyboard. Most likely, if the problem is here you'll have to replace the keyboard as a unit.

Before you give up and toss away the keyboard, check the continuity of the keyboard cable. It's unlikely that one of the wires inside is giving trouble, but it *is* possible. To check for continuity, remove the cable from both the keyboard and from the computer then set your meter to read ohms (x1). Touch the probes to the ends of each wire (pin 1 to pin 1, etc.) See Figure 6–21.

PRINTERS

There are many different printers available for the PC. Each has its own characteristics and construction. One printer might require a partial disassembly to just reset the switches. Another will have a built-in memory buffer. Still

FIG. 6–16 Testing the port for voltage. The black lead goes to pin 4. The voltage between pin 4 and all others should be between 2 and 5.5 volts dc.

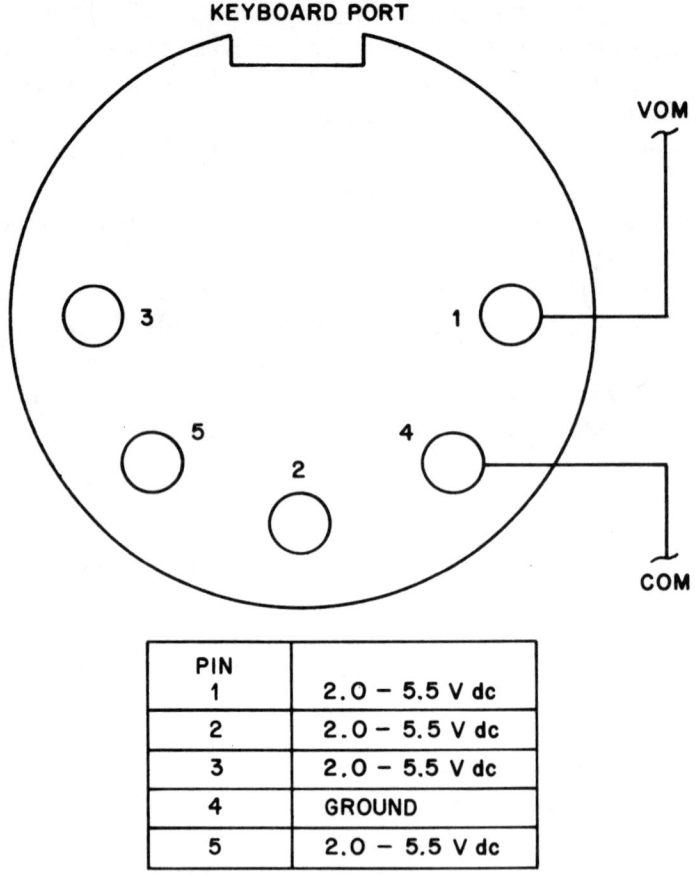

KEYBOARD PORT

VOM

COM

PIN	
1	2.0 – 5.5 V dc
2	2.0 – 5.5 V dc
3	2.0 – 5.5 V dc
4	GROUND
5	2.0 – 5.5 V dc

FIG. 6–17 Keyboard port pin locations.

another could have the dual capabilities of both computer printer and standard typewriter. Getting your printer to work may require special software patched to your regular programs. There are also differences in speed, print quality, and printing technologies. The printer may require a serial connection, a parallel connection, or may allow you to choose the one you prefer. With the wide range available, it isn't possible to give repair information on all makes and models.

The manual that came with your printer is the best source for specific information. Become familiar with it and find out what capabilities your printer has and how to take care of various problems that could come up.

PRINTERS IN GENERAL

Many printers give you the option of connecting it either as a parallel device or as a serial device. Parallel is the more common means of connecting a printer

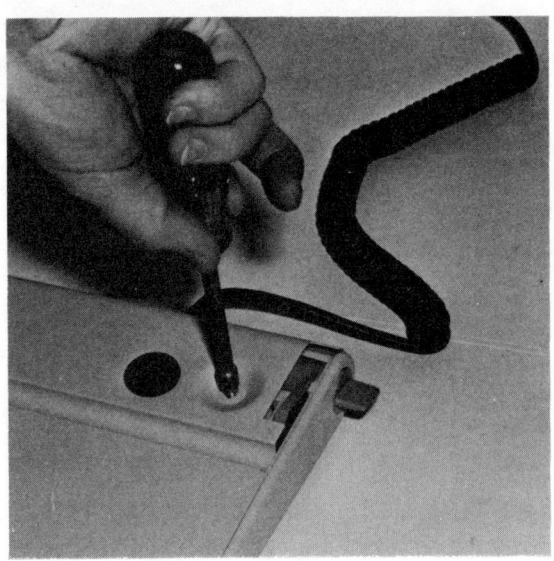

FIG. 6—18 To remove the cable from the keyboard to test for continuity, first remove the two screws on the bottom of the keyboard.

for several reasons. If the computer has one port of each type, the serial port is usually kept open for communications. Devices such as modems require a serial connection. By connecting the printer as a parallel device, the serial port is kept available.

Since the printer is a mechanical device, it is prone to more wear and tear than most things connected to the computer. It has at least two motors (for head and platen) and may have more. The print head moves back and forth across

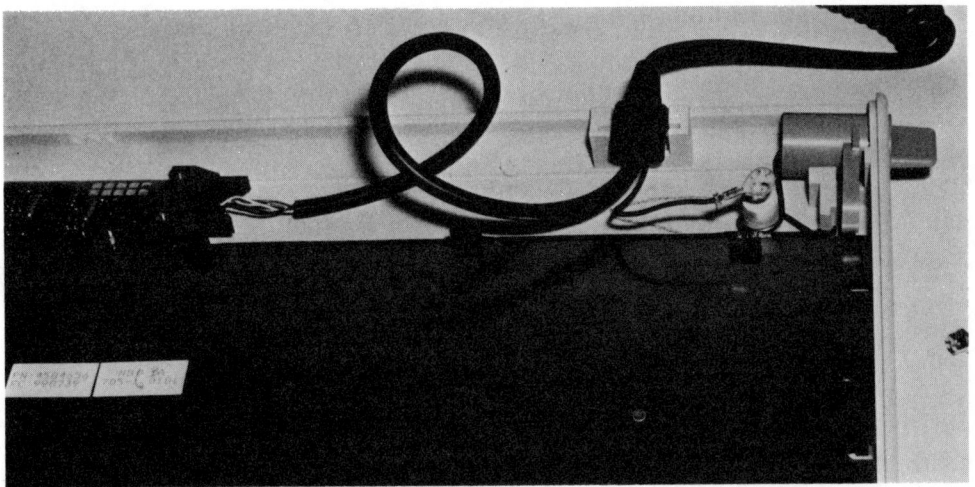

FIG. 6—19 The bottom of the keyboard is opened up for removal of the cable.

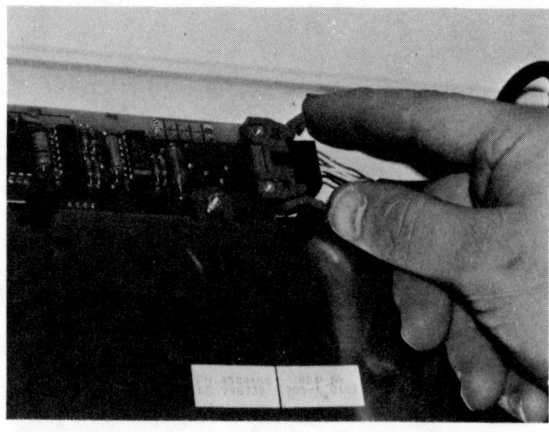

FIG. 6—20 Unplugging the keyboard cable.

the platen, and also either spins (as with a daisywheel printer) or has a print head that makes characters by punching at the paper with wires (dot matrix printer). All this motion causes wear. It can also create fair amounts of heat. If allowed to build up, heat can cause all sorts of damage, both mechanical and electronic.

The first thing to do is to get out the manual for your printer and get familiar with the information. Many manuals give specific error signals to let you know what has gone wrong. Also included will be information specific to your printer, such as how to remove the platen and other parts to free up a paper jam, how to load the ribbon, and so forth.

Paper can jam as it feeds through the printer. Even single sheets can cause a paper jam. Printers that use multiple sheets (with sheet loader or tractor) are even more prone to jams. Jams also tend to be more common if your printer is connected as a serial device. When a jam occurs, the printer can grind to a halt. Sometimes the jam isn't apparent. A few printers require a fairly complicated disassembly to get at it.

If the ribbon isn't installed properly, all sorts of strange things can happen. Part of a character might print, leaving the other part weak or nonexistent. It could shut down the printer entirely, or print a couple of characters and then act as though the signal had stopped.

The end of the ribbon is sensed by a small switch. As the ribbon reaches its end the switch stops the printer. This switch can also signal that the ribbon is used up if the cartridge isn't attached properly or if the switch fails.

Many printers have a safety switch in the lid. Lift the lid and the switch tells the printer to stop. It will tell the printer the same thing if the lid isn't closed all the way, or if the switch is faulty.

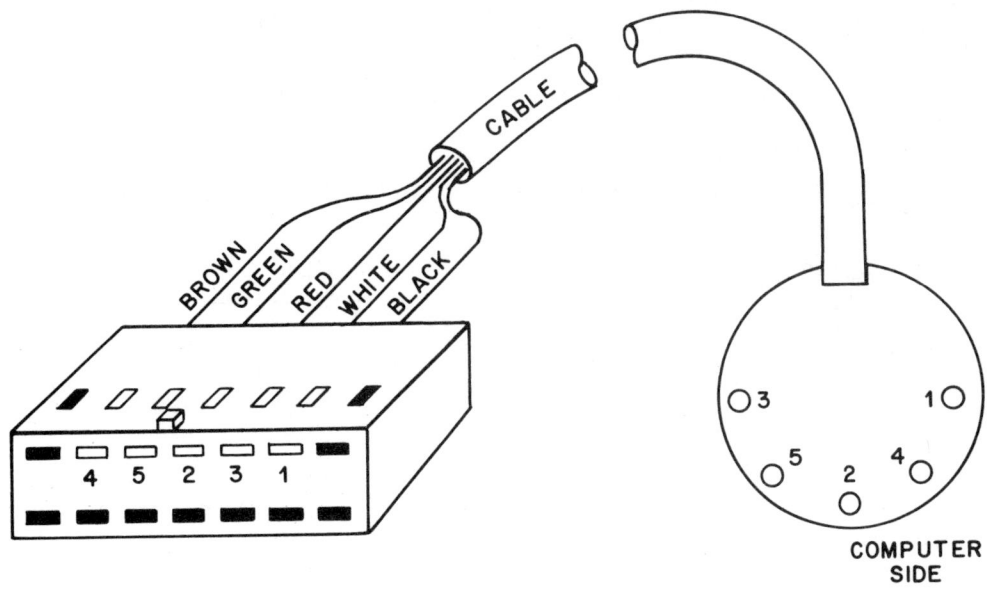

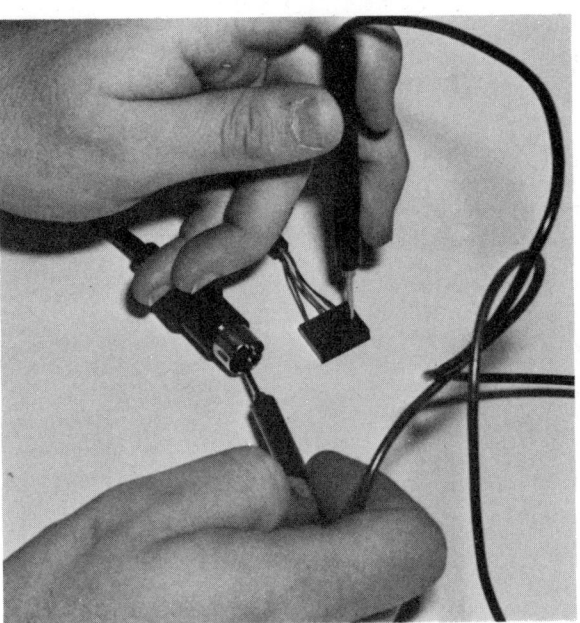

FIG. 6–21 Testing the keyboard cable for continuity. Touch pin 1 to pin 1, etc. (Do *not* use blackened-in switches.)

There are lots of adjustments possible with most printers. There are the usual spacing and forms thickness adjustments, the release catches, plus others. Just as a typewriter won't function properly if the adjustments aren't correct, neither will the computer's printer.

Checking for all of these things goes back to the standard rule, "Look for the obvious." The majority of the time the problem will be something very simple.

By performing occasional maintenance checks and cleaning the printer, you can greatly reduce the malfunctions. Clean the ribbon guides, print shield, and the inside of the machine. A build-up of ink or paper dust can cause problems. If your printer has a built-in self test, run it occasionally. (Run it once when you first get the machine and know that the printer is operating correctly. How else will you know what the results of the test are *supposed* to be?

This self test allows you to carry the diagnostics one step further. If the test shows that the printer is operating correctly, you'll know that the problem is in the printer interface, in the cable, or in the computer. You can eliminate the cable by testing for continuity with your meter. The diagnostics diskette can tell you if the printer adapter card and port are functioning correctly. About the only thing left is the interface in the printer itself.

PRINTER DIAGNOSIS

The error code for printer problems is 14xx. The first step is to disconnect the printer and run the diagnostics again. If the error code still appears, the problem is more likely to be inside the computer, and probably in the adapter card.

FIG. 6–22 A daisywheel can be cleaned carefully using alcohol and an old, clean toothbrush.

Williams: Repair & Maintenance of Your IBM PC (Chilton)

POWER SUPPLIES, KEYBOARDS, PRINTERS, AND MONITORS

If the error code disappears, you'll know that the fault is probably outside the computer.

If you suspect the printer, put it through the self-test cycle, if it has one. This should give you a good idea of where the problem is. If the self test operates, the printer itself is likely to be just fine.

Refer to the manual for the correct switch settings. If these aren't set correctly, the best the printer can do is punch out meaningless garbage. It is also important to have the software you are using "installed" for your printer, if the program calls for this.

If the printer is not making an impression remove the paper and look for indentations. The print head will make indentations in the paper if it is working, even if the ribbon is not.

If the printer is totally nonfunctional, the first things to check are all connections. The cable between the computer and the printer has 25 pins. Some are not connected to anything. This means that you'll have to take the cable heads apart to know the wiring. Inside, the wires are probably color coded. If even one wire is broken inside the cable, or if just one isn't making proper contact with the connector, the printer might refuse to function.

If there seems to be no power going to the printer, check the outlet, the power cord, and the fuse before assuming that the problem lies in the power supply. If power is obviously present, you can skip all power checks and go to the more detailed diagnostic steps, such as running the self test of the printer.

If power is getting to the printer but nothing happens, check once again all cables, connectors, switches, and the software itself. You can eliminate some things simply by knowing that the printer once functioned as it should.

MONITORS

Monitor problems are usually obvious. The screen may be blank. It may show an incorrect display. It could have an image that is tilted, too small, too large, out of focus, too dim, and so on. If the image on the screen is out of whack, the problem is probably with the monitor itself. If the problem isn't in the monitor, there are only a few other places it could be. You can find out quickly by running a few simple tests.

Tests, symptoms, and characteristics are slightly different for monochrome and color monitors. Others are identical for both systems. As always, begin with the simplest things first.

A client brought in a malfunctioning computer system. "The monitor isn't working," he said, "and I just bought it a few days ago." You should be asking yourself the paramount question now: "Has it ever worked?"

"No, but I tried everything I could think of. I know that the other monitor was working."

"Have you tried reconnecting the old monitor?"

The client did this and found that everything was working normally again. After a few more questions it was discovered that the old monitor had a monochrome display. The adapter card in the computer was for this monitor. The new display was color. No one had told the owner that there was a difference between the two, or that a different adapter card was needed for the color display to function.

It's easy to touch, bump, or otherwise change the various control knobs on the monitor. An example is the horizontal hold. If this knob is out of adjustment, the display could show nothing but a nonsensical pattern. The contrast or brightness controls might be out of adjustment, as could the color and tint knobs on a color monitor.

If no error code has come up on the screen, check all the knobs before you do anything else (including the on/off). You may have to turn them quite a bit to get any results. The proper adjustment may require changes in more than one control. (It is a good idea to maladjust your monitor when you know it is working correctly. This will teach you how to work the controls and will show what kinds of things can appear on the screen if the adjustments have been bumped.)

A sudden change in the display during operation indicates that something besides the manual adjustments is wrong. A simple adjustment probably won't help. Shifts during operation usually indicate a problem in the monitor itself, especially if no error code has come up.

If the self test finds a display problem, the error will be indicated by one long beep followed by two short beeps. Generally this means that the adapter card (expansion card) is faulty. The error code shown will be 4xx for monochrome and 5xx for color (assuming that the monitor can display the error code at all).

Note: The computer doesn't know what is being shown on the screen of your monitor. It relies on you to tell it. So when running the diagnostics diskette on the monitor you are better off doing it manually.

Check all cables and connections. (Obviously, you should know if the

---

Do not open the monitor except as a last resort, and even then only if you know what you're doing.

adapter card will actually support the kind of monitor you are using. If the monitor has ever worked properly, you'll know that the adapter card is the correct one.) If the connections seem solid, disconnect the cables and run the diagnostics again. If the error code is the same, you'll know that the problem is with the adapter card. If the error disappears, then the problem is either with the cables or with the monitor.

You can test the cables for continuity by using your meter set to read ohms in the x1 range. Place one lead on a pin on one side of the cable, with the other lead touching the same pin on the opposite side. The reading should show almost zero ohms. If it shows a large resistance (usually a reading near infinity), a wire inside the cable has broken and you'll have to replace that cable. Go from pin to pin until each wire in the cable has been tested.

Once you have eliminated the monitor and the cable, you'll have located the problem. If testing indicates that the monitor and cable are functioning properly, the fault is with the adapter card.

Before you go to the trouble and expense of replacing the card, check to see if it has been installed correctly and if there is any obvious damage, such as burned or loose components.

Is the card pushed all the way into the expansion slot? Try removing and reinserting the card to make sure that the contacts are solid. If you have a cleaner that will not leave a residue, clean the contacts on the board. (All lubricant type cleaners *do* leave a residue, so be sure to avoid them.) If you don't have a cleaner you can use the eraser of a pencil to clean the contacts. Be careful to brush away all the rubber tailings before bringing the board near the computer again.

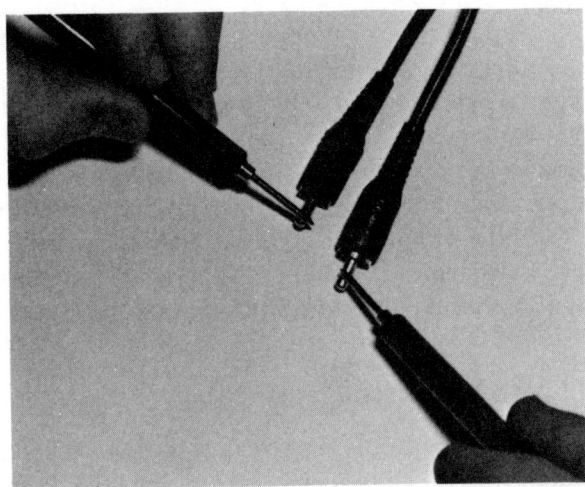

FIG. 6–23 Testing a monitor cable for continuity.

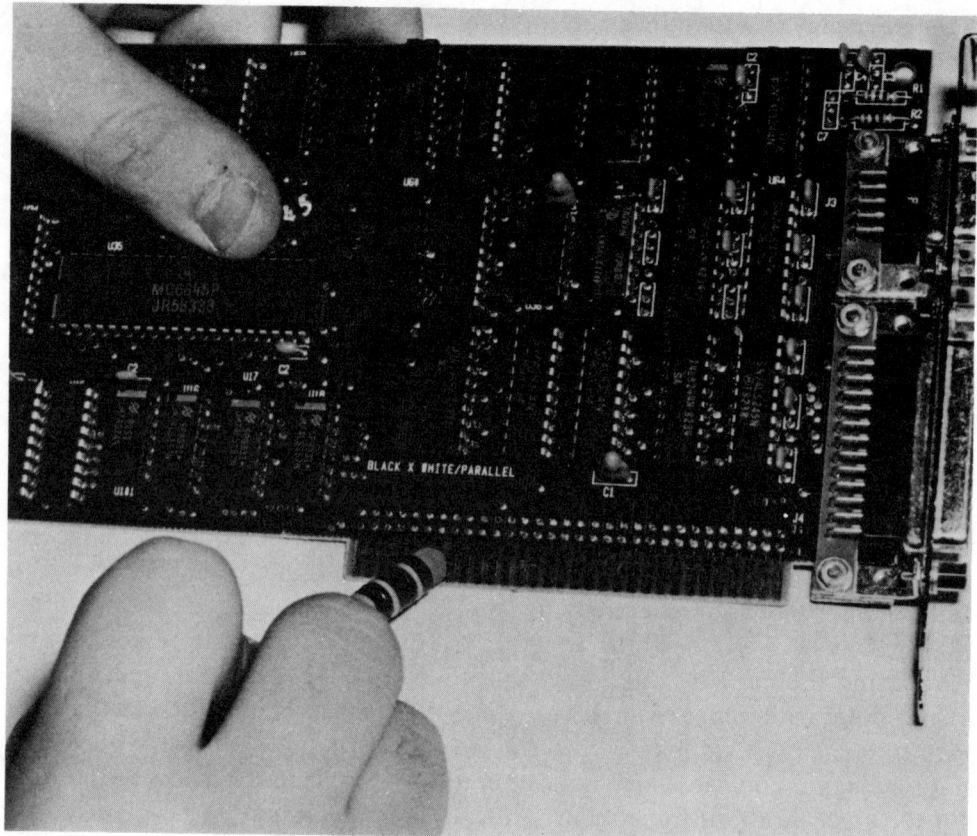

FIG. 6–24 Cleaning the board contacts with a pencil eraser.

A final check involves using another monitor, another cable, and perhaps another adapter card (one at a time in a process of elimination). If you don't have a friend with a PC, many dealers will be more than happy to help if they know that you'll be purchasing the replacement part(s) from them. (In this case, be fair to the dealer. He's in business to sell things, not to guide you to another "discount dealer." If he helps you find the problem, give him your business.)

If the problem still persists, the fault is probably in the adapter card. Repair is by replacement in almost all cases. Don't forget that the old expansion board will probably have a trade-in value.

BLANK MONITOR

If you have a problem so serious that there is no display at all, you can still run diagnostics to help locate the problem. To do this, you use the audio responses of the diagnostics diskette. This shows you the advantage of running the diagnostics when everything is working properly. Knowing what the correct responses are will let you run the diagnostics without seeing the display.

After you've eliminated all the obvious things as detailed above, insert the diagnostics diskette and turn on the power. When the program has loaded, press "O" (to tell the PC to start the diagnostics), then "Enter." Next press "Y" (to let the computer know that the options are correct) and "Enter." Press "O" (run tests one time) and "Enter." Press "Y" and "Enter" again (to take you through the typematic test for the keyboard). Each time you press the "Enter" key you should hear a beep.

If you have an IBM monochrome display, press "Y" and "Enter" three more times. If you have a color adapter card, even if it's driving a monochrome screen, press "Y" and "Enter" seven times. Then after the seventh time press any key eight times. Either way you should hear a beep after each "Enter" (there will be no beep the last eight times in the color adapter test).

At the end you should hear a long beep and a short beep. This tells you that the video portion of the test is over and that the diagnostics is about to test the drives. If you could see the screen, it would warn you to remove the diagnostics diskette and to replace it with a scratch diskette. Unless you have a reason to continue, shut off the power.

A lack of audio response anywhere in this sequence could mean that the system board (Chapter 5), the power supply (this chapter), or the connectors are faulty (you should have checked them earlier in this chapter). If the signals are all there and correct, the problem is more likely to be in the monitor or adapter.

SUMMARY

Power supply problems are not always obvious. The built-in protective circuitry will shut down the power if something is wrong elsewhere in the system. A "nothing happens" situation does not necessarily mean that the power supply is at fault.

Testing the power supply is a matter of eliminating devices one at a time, and of taking a few voltage measurements. Within just a few minutes you should be able to isolate the problem to something specific.

The keyboard is tested constantly. Not only does POST check it out every time you apply power, you test it yourself merely by using it. If pushing a key consistently does nothing, or if the key simply feels wrong, you'll know that it's time to perform a more thorough testing, or that it's time to replace the individual key or the keyboard itself.

Printers are famous for giving troubles. They are mechanical devices with at least two motors spinning along. Just connecting the printer in the first place can be a frustrating chore. Once it is up and working, you can keep it that way by some occasional maintenance, such as cleaning out the paper dust.

As with printers, a malfunction of the monitor is generally obvious. It also requires that you pay careful attention to what is happening, or what is not happening. The computer will test the monitor, but only to a certain extent. It can't tell if the monitor is displaying correctly. This is up to you. Learning what to expect from the monitor ahead of time will help you. Playing with the adjustments while the monitor is working properly can guide you later on if a problem comes up.

======================================

Preventing
Problems
7

The PC is so well designed and so well constructed that there isn't much to be done as far as maintenance goes. As with all computers, most of the activity takes place inside the circuits. There is very little to adjust, very little to clean, almost nothing to go wrong.

Even so, you can reduce repair costs and aggravation by performing a few simple maintenance checks now and then.

THE ENVIRONMENT

The most important aspect to the overall health of your computer system is its surroundings. The more dust and other contaminants that are around, the more often your system will give you trouble.

One owner used his computer to keep track of his electroplating company. The computer was kept separate from the plant, but not separate enough. The acidic fumes were obvious, both to the nose and to the computer. Every few months he would have a major computer failure. Twice in the first year he had to replace the drives, and once the entire system board.

Another company hired an operator who smoked heavily. Within a few months one of the drives was malfunctioning. When opened it was discovered to have a heavy layer of grime. The second drive was in nearly as poor condition.

The more pure and clean you keep the computer room, the better. You won't be able to keep it sterile and totally dust free, but this isn't really necessary. Your goal is to reduce the amount of contaminants.

When dusting around the computer, use a damp cloth. This will pick up and trap the dust, not just spread it around. It will help to keep dust out of the air.

While you're cleaning the area, clean the computer equipment. Dust will gather on the equipment, particularly on the monitor screen.

On the cloth, use a gentle cleaner or just plain water (squeeze out the cloth so it is just barely damp). Be especially careful with the monitor screen. Some have a nonglare coating that can be damaged. *Do not* have open containers near the computer! It's too easy for liquids, dusts, or whatever to spill. When they do, they have the unfortunate tendency to fall exactly where they can do the most damage. (Murphy's Law at work.)

Be very careful about getting liquids of any kind into the electronics of the computer. A slightly damp cloth can be used to clean the cabinet, monitor, printer case, and the outer edges of the keyboard. *Do not* use a damp cloth inside anything electronic. You're asking for trouble if you do. Dust will have very little effect on the electronic components under normal use. For example, it won't hurt the system board. But it *will* hurt anything mechanical, the read/write heads, the diskettes, etc., so keep dust to a minimum.

The inside can be cleaned with a vacuum cleaner (with a soft brush) if the dust has built up. This isn't really necessary, as the components inside the computer itself aren't sensitive to dust as long as it is dry. The internal parts of the printer are more important. Paper gives off a surprising amount of dust. This paper dust can collect in all the wrong places and can jam the mechanical parts.

Use a cloth that is just barely damp for cleaning around the computer equipment. Don't use anything like a feather duster. This applies to the entire area around the computer. You don't want to be throwing dust into the air that settled on a shelf.

Even when cleaning the floors, be cautious. If you use a vacuum cleaner, move it slowly to keep down the amount of dust. If you have a floor that gets mopped, don't let it get too wet. This will increase the humidity in the room.

There is no such thing as being *too* clean around a computer. Keep the area as clean as possible. But when the area is cleaned, be sure that the cleaning doesn't just toss all the dust and grime into the air. The computer circulates air inside it for cooling purposes. Some of this is sucked in through the disk drives—the one spot where dust can do the most damage.

Williams: Repair & Maintenance of Your IBM PC (Chilton)

PREVENTING PROBLEMS

DRIVE HEAD CLEANING

There are two opinions about head cleaning. One is that it is never necessary and can only cause damage. The other is that you should clean the heads after every so many hours of use and that the cleaning will cause no damage at all.

Both are true, sort of. Obviously, you're better off leaving the heads alone as much as possible. But, when cleaning is necessary, it has to be done. Nor can you afford to wait until the recorded data is full of errors due to a dirty head.

Anything abrasive used on the heads is obviously capable of damaging the heads. The same goes for cleaning fluids that can affect other parts in the drive. If you stay with a well-known brand you should have no trouble at all.

The idea of cleaning the heads after so many hours can also be misleading. The key is your own environment. Obviously if you work in a sterile environment and use only the best-quality diskettes, deposits will be minimal. You won't need to clean the heads very often. On the other hand, if you're a heavy smoker or have the computer in a poor air environment (shame on you!), you will have to clean the heads more often.

Head cleaners can only take off "new" deposits. If you let those deposits build up over a long period of time, they become permanent parts of the heads. No cleaning kit in the world will take off such deposits, at least not without destroying the head at the same time.

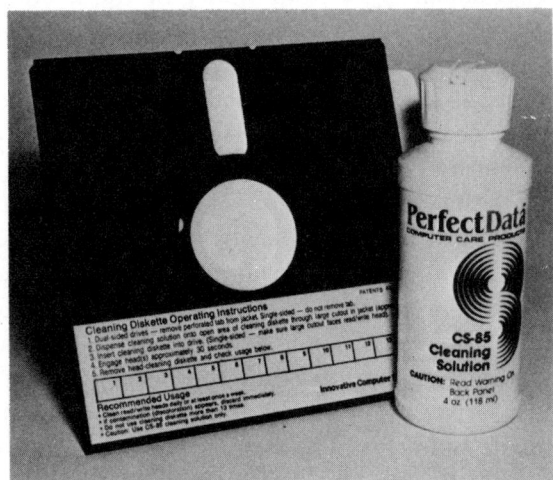

FIG. 7–1 Drive head cleaning kit.

There are various ways of cleaning the heads. Before people realized how much damage could be caused, abrasive cleaners were used. These literally scratched away the contaminants. Fortunately, very few of these are around now. Such a cleaner will certainly take away all the built-up grime. It will also take away the surface of the head.

The next step is a "nonabrasive" head cleaner with a bottle of fluid in the package. Generally such a package is less expensive and also tends to last longer. Again you have to be careful of having a rotating pad that is too abrasive. (Stick with a reputable brand and you won't have much to worry about in this respect.)

Generally, the user tends to either oversoak or to undersoak the pad. Both can cause problems. If the pad is undersoaked, the abrasive action is increased. (It also won't clean the head as well.) If it is made too wet, you'll have the excess fluid sloshing around inside the delicate parts of the drive. It will evaporate before too long, but in the meantime it can cause problems. Worse yet, there's no way to know if harmful materials have been deposited elsewhere in the drive or other damage has been caused that will show up at a critical moment.

The most expensive head cleaners are presoaked with cleaner. Carefully measured amounts are already on the diskette cleaning pad, which means that you have virtually no risk of having the cleaning fluid (and the stuff it has dissolved) dripping into places you don't want it.

The disadvantage of such a cleaning kit is that the cleaning diskette not only costs more to begin with, it also wears out sooner. Since there is never an excess of cleaning fluid, the fluid will evaporate more quickly from the cleaning pad, making the diskette useless, if not dangerous. The advantage is that you have much less risk of damaging the heads, the drive itself, or other diskettes you put in. Personally, I believe that this lower risk easily warrants the higher cost. The difference in price just isn't all that much.

You can clean the heads manually, of course. This involves some work on your part, and a certain degree of risk. To do this you'll have to remove the drive and the circuit board. You simply can't access the heads safely otherwise. Each time you do this, you take up your own valuable time and risk damage to the circuit board. Generally this is *not* the way to go about it.

If you choose this method you'll need some very clean cotton swabs (the tighter the better) and cleaning fluid. Isopropyl alcohol will do, but *only* if it is pure. Much of the alcohol available to the general public has water and certain oils in it. Some has other contaminants that could be harmful. Using alcohol that is available "off the shelf" is often a great way to guarantee damage to the heads and drive. Technical grade alcohol is available through chemical supply companies. Be sure that you specify isopropyl alcohol with no contaminants.

You can also use fluid head cleaner for audio tape players. Again, it is important that you get the very best possible. *Don't* try to save a few pennies. The heads of your disk drives are too important, and too expensive to replace.

DISKETTES AND SOFTWARE

The more you use a diskette, the more chance there is of trouble. In Chapter 3 you learned how tough, and how delicate, a diskette is. The best-made diskette can still present problems.

The major problem is that of the "allocation table" or "directory track." Each time you use the diskette, the computer seeks out this track to learn what you have on the diskette. This means that the life of the diskette is directly proportional to the number of times you use this track.

The second major problem is that of editing. Changes in a diskette file can scatter that file all over the diskette. A disjointed file is much more prone to hand out errors than one that is in a logical sequence.

Think of it as several very long letters to friends, and that you write these over a period of a year or more. Before you send them you want everything to be just right. So you go back to the loose pages and add little notes. Over a period of time you have a massive stack of scribbles and pages. It would be almost impossible for you to keep track of which pages go to which friends.

The computer doesn't have quite this much trouble, but it still has to keep track of which changes you've made, and to which files. Those changes may end up scattered all over the diskette, and it would be like having a heavy wind get hold of those letter files.

The solution is to make copies often. You have two choices. The "Diskcopy" routine will make an *exact* duplicate of the original. This method is fast and will copy exactly what is on the original, including the "blank" temporary files used by some programs and your edited "scattered" files. The "Copy" routine is slower and will ignore those temporary files. This command has the advantage in that it will take a disjointed original and arrange it on the copy diskette in the correct order. Fragmented files are thus restructured.

Instructions for using "Diskcopy" and "Copy" are in your DOS manual. If you don't already know how to use these commands easily, learn!

Programs used often should also be copied on a regular basis, if the program allows this. Many don't in order to thwart software pirates. The new copy also has a new allocation track, which will bring a worn program back to life again.

Many "Copy Protected" programs can be copied with the use of a special program, such as COPYII PC. The purpose of these programs is not to help software pirates but to allow legitimate users to back-up their important (and expensive) programs. After using one of these programs, run the new copy immediately to make sure that it works.

Don't wait until the program or data diskette has failed before making the back-up. Do so immediately! Right now, in fact. If you have diskettes that are used often that haven't been backed up recently, put down this book and do it—NOW! If the back-up sits on the shelf or in the box for a year, great. You don't make the back-up to use—you make it to ensure that a diskette failure doesn't knock you down and out. (Wouldn't it be great if you had a back-up heart? Back-up software is nearly as important.)

DIAGNOSTICS

If you haven't already done so, read through Chapter 2 in this book and read the diagnostics section of your *Guide to Operations* manual. Become familiar with what is possible, and what is not.

Assuming that your computer system is functioning properly, run the diagnostics program right away. Become familiar with how things are *supposed* to look. If something goes wrong later on, you'll know better how to track it down.

It's a good idea to run the diagnostics diskette on a regular basis. How often will depend on your use of the computer. Once each month will probably be just fine. (I use my own PC heavily and the once per month routine serves me well.)

The few minutes spent in letting the computer test itself thoroughly is a worthwhile investment. It also serves to remind you how things should be. If you wait a year or more before running the diagnostics, you could easily forget. Even the notes you've made (you *have* made notes, haven't you?) may not make much sense a year from now.

If you can afford the investment, the advanced diagnostics diskette available from IBM (through your dealer) does a much better job of keeping you informed of possible troubles. It has certain test routines that are not available on the standard diagnostics diskette. The accompanying manual is full of helpful tables and charts and contains a wealth of other information. The cost is $150 (at time of this writing), which makes it an investment that the average person won't care to make.

At $40 the Verbatim Disk Drive Analyzer program mentioned in Chapter 4 is an excellent and a relatively inexpensive means of keeping a close eye on the most critical part of your system—the drives. Run this program on a regular basis and you'll spot most common problems in the drives before they do serious, irreversible damage.

OTHER STEPS

It's all too easy to get used to the power of the computer and forget that it is quite delicate in certain ways. Each hour you spend with flawless operation tends to make you forget that things *can* go wrong.

Make it a habit each time you sit down to the computer to review in your mind all the things that can go wrong. Pay close attention to what it is doing, and how things are working. The more you use a particular program or function, the more important this is.

If you have a friend with a PC, sit down and use your friend's machine occasionally, with your programs and data diskettes. Trade the opportunity, and give your friend a session with your computer under the same conditions. This will tell both of you quite a bit about how well the computer system is functioning.

For example, if a diskette or program functions perfectly on your machine and not on the other, something is wrong with one of the two. Don't assume that the problem is with your friend's PC. If the heads in your drive are misaligned, you'll be recording information that can only be read on a drive with heads that are misaligned in exactly the same way.

Such a swapping of computer time will help to keep you informed of any serious malfunctions that are occurring, especially if you do it on a regular basis. It gives you a basis for comparison that many people don't have.

If you don't have a friend with a PC, contact a dealer. Chances are he'll let

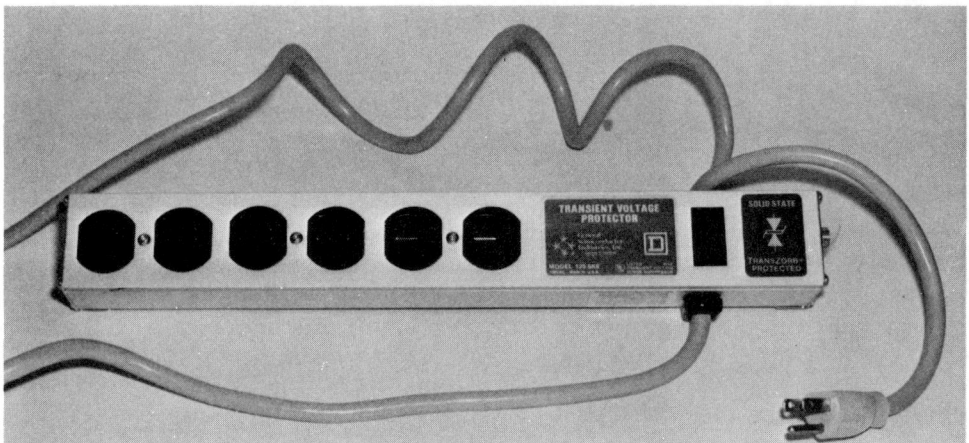

FIG. 7–2 A line protecting device will prevent power surges from damaging your system.

Williams: Repair & Maintenance of Your IBM PC (Chilton)

MAINTENANCE ROUTINE

Daily
1. Make back-up copies of all data diskettes you've been using. This should be done at the end of each session and periodically during the session.

Weekly
1. Give the computer area a quick cleaning to cut down the amount of dust.
2. Perform a "Copy" routine on any data diskettes that have been used and edited heavily.
3. Run Disk Drive Analyzer, if you have one.

Monthly
1. Thoroughly clean the entire area, including the inside of the printer.
2. Clean disk drive heads.
3. Test devices and equipment that are rarely used.

Occasionally
1. Test back-ups already made.
2. Make new back-ups of important programs and data if needed.
3. Run diagnostics (even if nothing has malfunctioned).
4. Spend some time with your diskettes on someone else's PC.
5. Learn something new about your system and its programs.

The actual frequency of the maintenance routines will be determined by your use of the system. For example, if you only work at the computer a few hours per week, you probably won't need to clean the drive heads as often. This listing is only a general guide. Set up a schedule appropriate to your own circumstances.

you make use of one of the floor machines for a short time, especially if you've been a good customer.

Lacking this, consider joining a PC users group. Other members in the group are likely to be after the same sort of cross checking. As you check your system, they are checking theirs. (You'll also end up with a valuable source of information, and plenty of new friends with your same interests.)

SUMMARY

There is very little maintenance to be done with your computer system. Invest just a few minutes per week and you've done just about everything necessary to keep your system running without trouble.

Williams: Repair & Maintenance of Your IBM PC (Chilton)

The environment around the system is critical. The cleaner you keep the general area, the fewer problems you'll have, and the less maintenance you'll need. Keep dust and other contaminants away from the computer as much as possible.

If you ignore the drive heads for too long they could develop a permanent build-up of particles that no cleaner can remove. The result will be faulty reading and writing of data, at unpredictable times. A periodic cleaning of the heads, using the best-quality head cleaning kit, will help to keep the heads working perfectly for many years. How often you clean the heads will be determined by how much your computer is used and what the surroundings are like.

The most delicate part of a computer system is the software. Handle the diskettes carefully. Just as important, make back-ups of all diskettes as a regular part of your work schedule. The cost of diskettes is cheap compared to the cost of replacing lost programs or data.

Every so often, run the diagnostics diskette. It's unlikely that you will find any problems, but doing this step will help to spot malfunctions before they become serious. It will also help to remind you of how the diagnostics *should* perform.

==================================

Adding to Your System
8

Just when you think you have everything you need as far as computers go, something else comes along. The printer you have may not be printing fast enough for your needs, or perhaps the character quality needs to be improved. If you bought the computer with just one drive, you're almost guaranteed to want a second drive very soon. That monochrome display might be replaced by one that puts out pretty colors. You're certain to be adding new software programs from time to time.

Even if you don't add something to your system, the day could easily come when an existing piece of equipment requires replacement. The steps in replacing a device are nearly the same as adding that device for the first time.

This chapter will show you how to handle some of the most common additions. Your own circumstances may be slightly different, depending on the equipment you're using. However, the general guidelines presented here should help.

It's important that you become as familiar with the new device as possible. If it has an installation manual with it (most accessories do), read it through carefully, cover to cover. Don't skim or skip. Certainly don't try to install the device while reading the manual for the first time.

The biggest problem that most people run into is that of being in too much of a hurry. That new memory board you've been saving up for arrives, and seems to cry out for immediate use. Off comes the cabinet, and in goes the board

with little more than a cursory glance at the installation manual. The result—
the thing doesn't work.

Slow down. And then go slower yet. Take your time. If you don't understand the instructions, go back and read them again. Have a clear idea of what
you're doing before you start and the installation is much more likely to work.

You'll actually save time (and much frustration) by taking the time to
become familiar with the information first.

The *Guide to Operations* manual that came with your computer is a fair
guide for many installations. In it are lots of tips and instructions. Even those
sections that don't seem to apply to what you're doing can be of help. If you
haven't already done so, take out this manual and read through it. It doesn't
matter if the accessories you have were manufactured by someone other than
IBM. Much of the information in the *Guide to Operations* applies to all additions you'll make.

Before actually installing anything, go through the procedure at least once
mentally. Do you know which steps to take and when?

Check and double check any switch settings you've had to make, both in
the computer and in the new device. If no switches are to be changed, be sure
that you know this as well.

Finally, *never* install a device while there is power flowing. Shut off the
switch. There is rarely a need to unplug the computer. In fact having it plugged
in provides a safety by grounding everything. If you leave the computer plugged
in, be aware that the line and power switch are still "hot."

A SECOND DRIVE

If you've been trying to operate your computer with just one drive, you'll know
how severely limited you are. Not many people will be satisfied with just one
drive for very long. Few even buy a computer with a single drive. (At least this
is true of people buying the PC.)

Single-sided drives also present limitations that soon become tiresome.
There are so many advantages to the double-sided drives, and the additional
cost is so small, that it's surprising that anyone has the single-sided versions any
more.

Installation for single-sided or double-sided drives are the same. The computer will recognize the presence (or absence) of the second read/write head
simply by the way the drive is assembled. Diagnostics can also recognize which
kind of drive you're using.

The difference in installation comes in whether the drive is installed as
being drive A or drive B. Installing a drive as drive B requires a minor modi-

FIG. 8–1 Half-height drive.

fication to the drive circuit board. This is easy to do and takes about 30 seconds. (It's possible that the drive to be used as drive B will come to you ready to go. Check it just to be sure.)

At the rear right of the circuit board on top of the drive is a terminating resistor package (see Figures 8–2 and 8–3 for exact location). It looks just like an IC module, so be sure that you know which chip package you're working with. This terminating resistor tells the computer which is A and which is B. If both drives have the package in place, the computer is being told two things at once. Remove the terminating resistor package from drive B.

Many drives also have a jumper block that must be modified when installing a new drive. This socket is near to the terminating resistor. Certain pairs must be "jumpered," or connected. The easiest way to do this is to check the jumpers on drive A. The same pins will be jumpered on drive B. (If you're replacing a drive, simply check the locations of the jumpers on the old drive.)

Generally there is only one set of pins to be jumpered. And usually this is the third pair from the rear of the drive. A clipped lead from a resistor or other component will work fine. In a pinch, a staple does just fine. Bend the leads carefully and be sure that you're connecting the correct pair.

Although this is the normal procedure, these instructions are only a general guide for your reference. Check the instructions that come with the new drive. If you're using a drive other than Tandon or CDC, the jumpered connections could be different.

When installing a second drive to your system, the computer has to know which drive is A and which is B. Part of this is taken care of by the terminating resistor, as mentioned above. Another important part is accomplished by twisting (or reversing) some of the wires in the signal cable. (Chapter 4 shows you which wires are twisted.) This normally won't be a problem, since the IBM cable already has the wires reversed. You'll only have to worry about it if you try to make your own ribbon cable. All you have to be concerned about is that the twisted side of the cable goes to drive A.

To do the physical installation of a second drive, remove the plastic plate covering the hole where the drive is to go. This is easily done. Remove the cabinet. You'll see two metal fasteners on the rear of the cover plate. They come off easily, after which the cover plate comes loose. (See Figures 8–5 and 8–6.)

FIG. 8–2   Location of terminating resistor package, Tandon drive.

Williams: Repair & Maintenance of Your IBM PC (Chilton)

FIG. 8–3 Location of terminal resistor package, CDC drive.

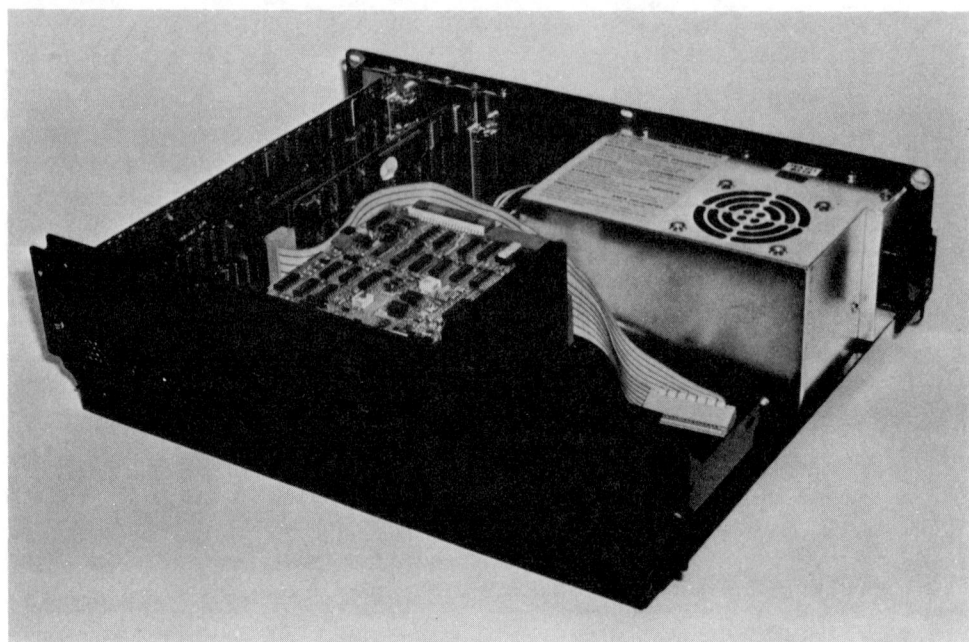

FIG. 8–4 The PC with one drive installed and ready for the second drive.

Williams: Repair & Maintenance of Your IBM PC (Chilton)

ADDING TO YOUR SYSTEM

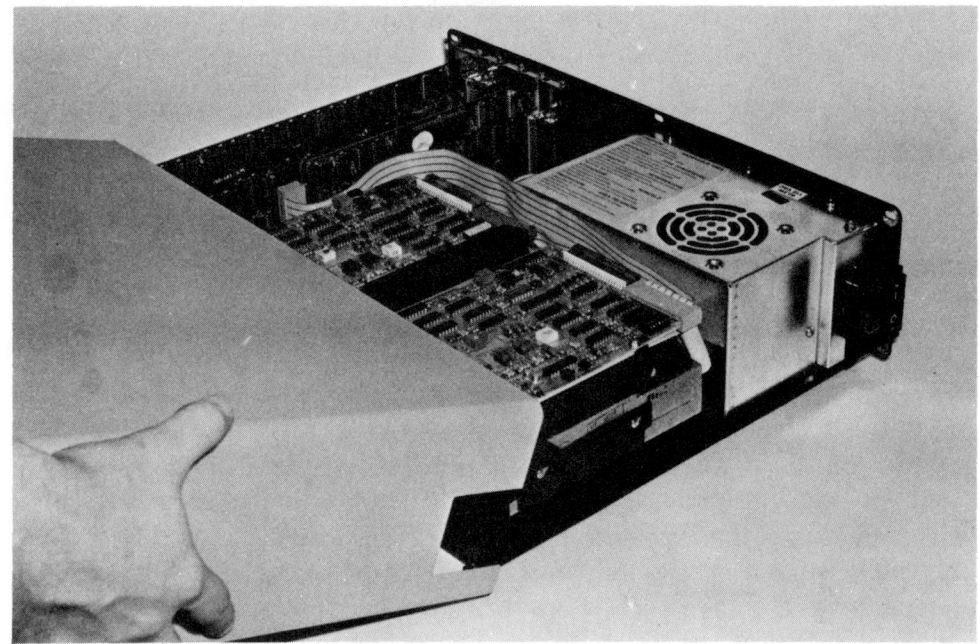

FIG. 8–5   To install a drive, first remove outer case.

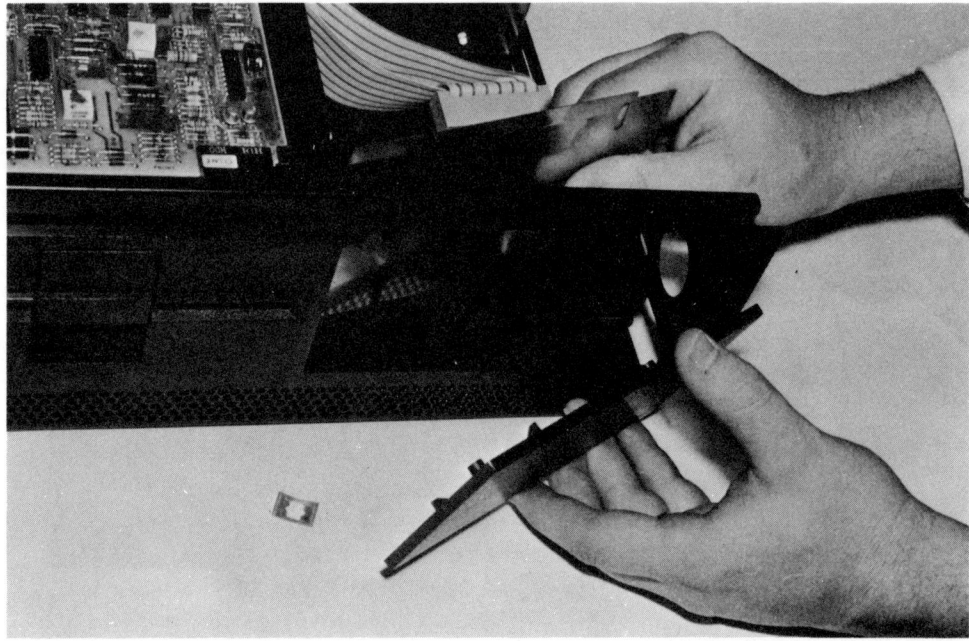

FIG. 8–6   Remove plastic drive opening cover.

Williams: Repair & Maintenance of Your IBM PC (Chilton)

Insert the drive about half way into the hole (Figure 8–7). This gives you room to make the connections with the power supply and signal cable (Figures 8–8 and 8–9). Both cables are keyed and attached only one way. Inspect the connectors and see how they are keyed.

The power supply cable will have two of the corners of its connector angled and the other two squared. This cable will mate with the drive connector stiffly, so be sure to plug it in all the way or it might fall out. (If this happens the drive motor will not turn and the drive LED will not light. If these are intermittent, check this connector before anything else.)

The signal cable (the flat ribbon cable) will have a plastic insert on one side of the connector which mates with a slit in the edge of the card of the drive.

After connecting the cables, push the drive in, but not all the way. Use a few business cards as spacers between the metal case and the lip of the disk drive (Figure 8–10). The idea is to leave a little space between the front of the drive and the metal system unit cabinet. Then, if the computer is picked up or moved, the flex in the cabinet won't break the plastic plate.

The drive is secured with two screws (Figure 8–11). Use only the screws provided with the drive and screws them in. The screws should be tight, but not overly so. Replace the cover and you're set to go.

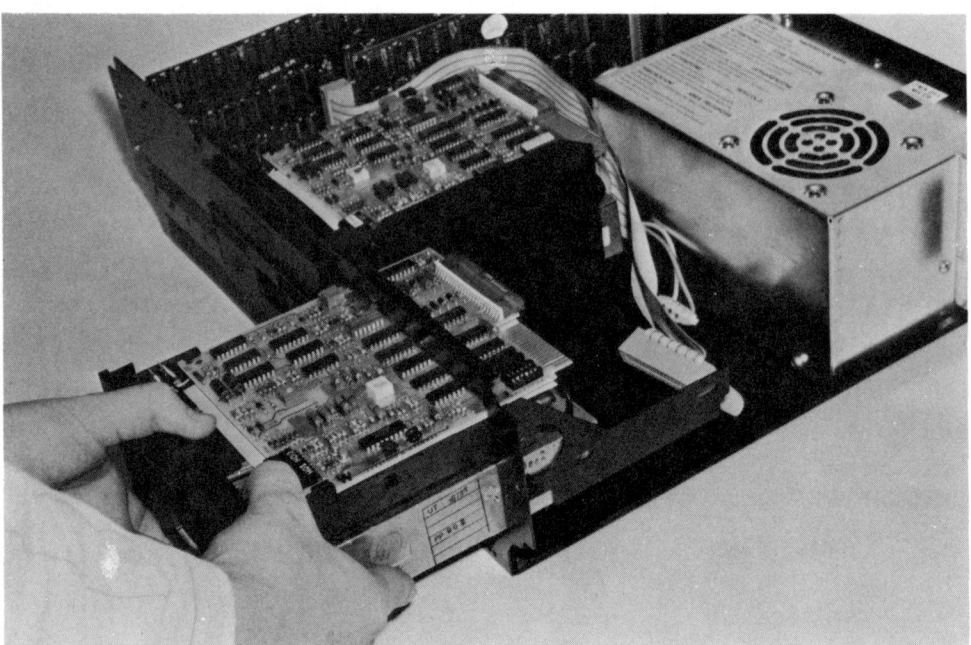

FIG. 8–7 Slide new drive in about half way.

Williams: Repair & Maintenance of Your IBM PC (Chilton)

ADDING TO YOUR SYSTEM

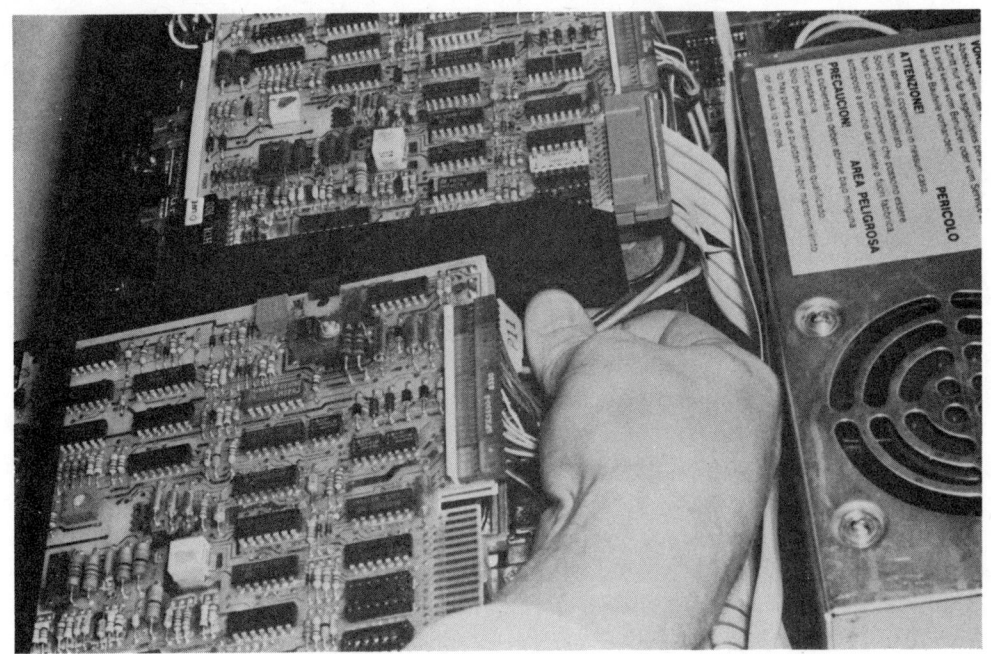

FIG. 8–8 Connect power cable.

FIG. 8–9 Connect signal cable.

Williams: Repair & Maintenance of Your IBM PC (Chilton)

ADDING TO YOUR SYSTEM

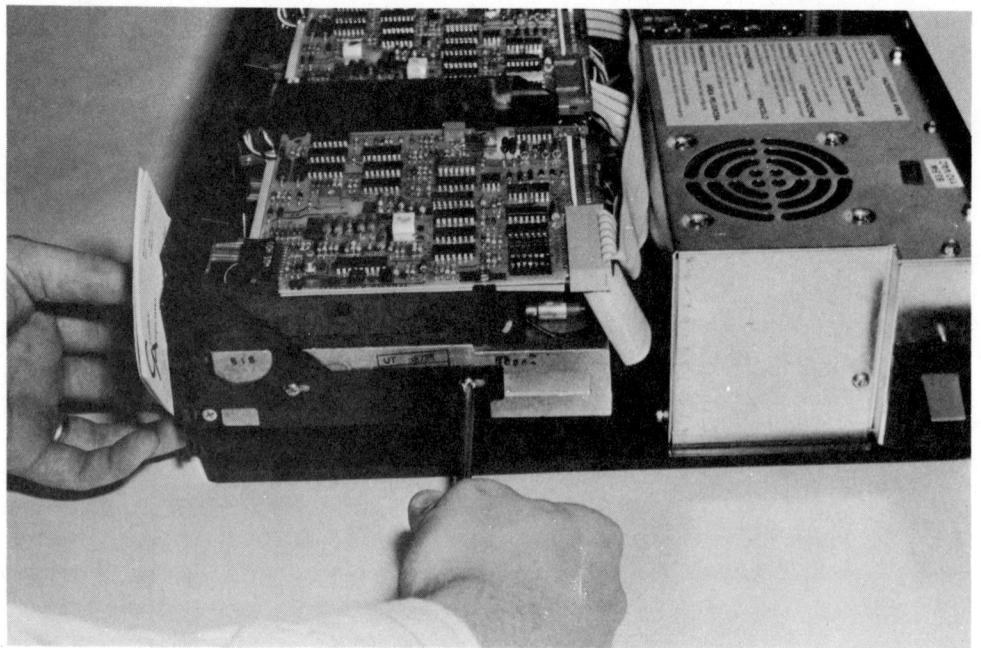

FIG. 8–10 Slide drive in, using a few cards as spacers. (Remove the cards afterwards.)

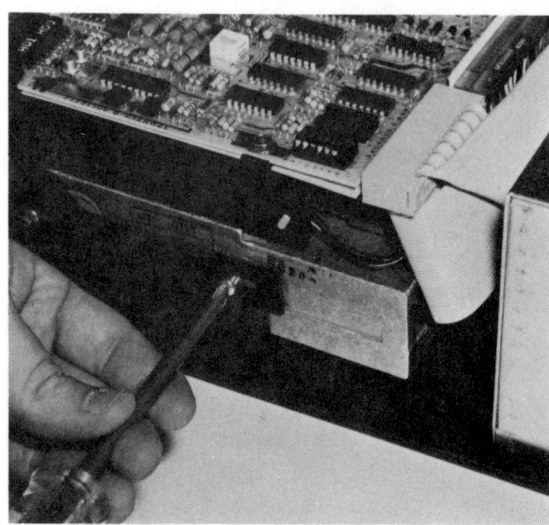

FIG. 8–11 Tighten holding screws.

The IBM floppy drive controller card is capable of handling up to four disk drives. Normally you use only two, which are installed in the main cabinet. If you require more floppy drives, these will probably be mounted in the expansion chassis. Normally you'll use a second drive adapter card (in the expansion chassis), but it is possible to have the drives connected to the main chassis through the 37 pin d-shell connector (external). This gives them the power supply that they need. Unfortunately, IBM does not make the cable you will need to operate the two additional floppies. Although you could make a custom cable, you would probably find it easier to install another drive controller card in the expansion chassis.

PRINTERS

Printers come in two basic types and in two basic configurations. The two types are dot matrix and those that use a wheel or ball (letter-quality printers). The

FIG. 8–12 Graphics printer.

Williams: Repair & Maintenance of Your IBM PC (Chilton)

FIG. 8–13 Inside a printer.

two configurations are parallel and serial. A serial printer uses data that is sent one bit at a time. The parallel printer uses data that is sent one byte (eight bits) at a time. Each has advantages and disadvantages. Most people use the parallel configuration since this leaves open the serial (COM) connectors for other functions, such as modem communications.

Cabling is the biggest consideration when installing a printer. If you try to build your own cable, it can get very complicated. Fortunately, the IBM PC is so popular that finding the proper cable for any printer should be easy.

Before even attempting to make any connections, get out the manual for your printer and read it cover to cover. Become familiar with the operations, options, switch settings, and so forth. The more you know, the better.

Visually inspect the printer. Locate the controls and learn how to use them, even if you don't think you'll need those options. Also look at the print head. Some are tied in place to reduce damage in shipping. Others may be packed along the sides.

As with many computer devices, printers usually have DIP switches to set. The printer was designed and built to accommodate a variety of computers. The switches allow you to configure the printer to your needs. The switches may also set other functions of the printer. Don't forget to shut down the power

Williams: Repair & Maintenance of Your IBM PC (Chilton)

ADDING TO YOUR SYSTEM

before changing any of the DIP switch settings. Most printers won't recognize switch setting changes while power flows. Others can be permanently damaged.

The manual that came with the printer should tell you everything you need to know about the switch settings. Read it carefully. You can't damage anything by not having the correct settings, but the operation of the printer won't be right, if it works at all.

HARD DRIVE

There are many "Winchester" drives on the market for the IBM PC. The most popular is the ten-megabyte model that comes standard with the PC-XT. This same hard drive is available from IBM for the PC. The easiest way to install it would be to purchase the expansion chassis. This piece of equipment not only gives you a ten-megabyte Winchester, but also the power supply needed by the

FIG. 8—14   Hard (Winchester) drive.

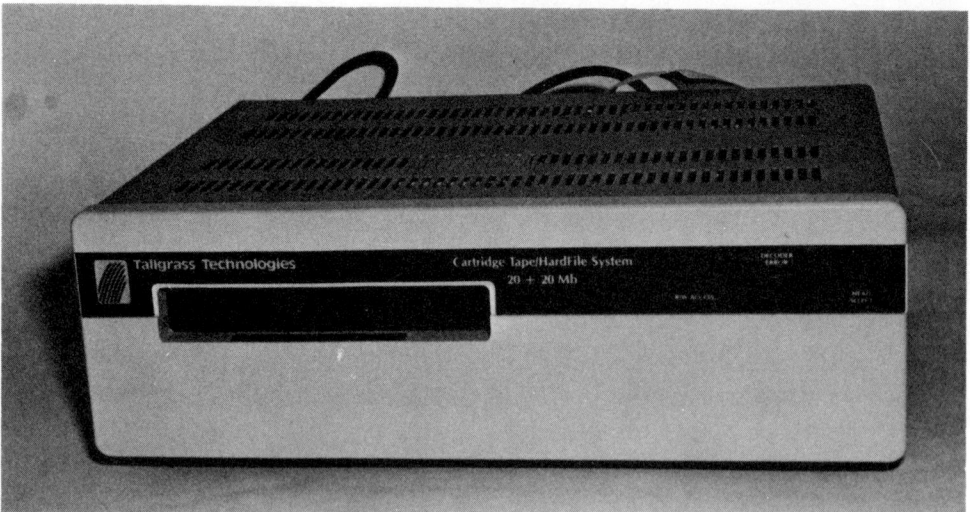

FIG. 8–15 Tape back-up for hard drive.

hard drive. (The cabinet for the expansion chassis also matches perfectly with the original system unit cabinet.)

This expansion chassis has no system board but rather an expansion "bus" which allows you to install six more options. There are openings in it for two drives, just as with the system unit cabinet, which allows you to put in not one but two hard drives, if you wish. It contains a heavier duty power supply (150 watt compared to the 60-watt power supply of the main unit) for the added power consumption of the hard drive.

As with almost everything else, there are a number of manufacturers who make hard drives. Each is different and requires different commands. Some are only marginally compatible with the PC. Do your research carefully when shopping for a hard drive!

For example, most drives will have a back-up utility (a means of storing the hard disk files onto floppies for data safety). Some support the use of special cassettes as back-up. The commands required could be completely different from what is stated in the PC-DOS manual, so be sure to find out how to back-up your programs and data from the dealer. (The same is also true with the "format" routines.)

There are two basic methods of backing up data and programs for a hard drive. One is to simply use the "Copy" command from DOS and place the programs and data on floppies for safety. This can be tedious. You will have to copy each file individually, since the "Copy *.*" will take only the first files off

Williams: Repair & Maintenance of Your IBM PC (Chilton)

ADDING TO YOUR SYSTEM

the disk. (As soon as the floppy is filled, the routine will stop.) The same problem exists when using the "Diskcopy" command from DOS. You simply can't copy the contents of a ten-megabyte hard drive onto a 360K floppy.

If you're going to back up your hard drive with floppies, number each of the diskettes and label them carefully. This way, if you ever do need to restore a crashed hard drive, you'll know exactly what was in there, and in what order.

The second method of back-up is to use a special cassette. This method has the advantage in that the entire contents of the drive can be placed on a single cassette, instead of on a number of diskettes. It has two major disadvantages. The first is the additional cost of purchasing and connecting the back-up machine. This isn't much of an expense when you consider the value of the data and programs stored on the hard drive, but it is one more item that has to be purchased. Second, and more important, there is no way to check the integrity of the back-up copy on some systems unless you erase everything on the hard drive and try to reload from the back-up cassette. (This may also be true of backing up your programs and data files on floppies.) Trouble is, by the time you find out that the back-up is no good, the original is gone.

A customer bought a new hard drive, the cassette back-up, and all the things that go with them. He dutifully backed up everything in triplicate, including a massive 2.5-megabyte data file for his business. One day the hard drive crashed and refused to boot. All the data on it disappeared into electronic limbo.

No problem, he thought. He got the hard drive repaired and went to restore the data from the back-ups. Then he found out that the back-up program had malfunctioned and that the back-up tapes were blank. It was a software problem, not mechanical, and was easy to fix—but that did nothing at all to restore the lost data. (Generally the tape back-up is considered to be one of the most efficient and safest methods.)

You can reduce the risk by using only the highest-quality diskettes, cassettes, and machines for back-up (and the best-quality hard drive in the first place, of course). If you purchase the entire package from a local dealer, he may be willing to handle the installation of your programs, especially the tricky ones. He should also check to make sure that everything is functioning. If you have to wait another few days to get these services performed, wait. It's better than finding out the hard way that something isn't working right.

When you make your own back-ups, monitor every step of the procedure carefully. In most cases, the program will tell you if it is having troubles with the back-up. Then you can take the proper steps to correct the situation.

The installation of the hard disk hardware is very simple and requires nothing more than inserting it into the drive hole in the chassis and plugging in the connectors. An external hard drive is a little more complicated, but is still not difficult.

The software installation, unfortunately, is not all that simple. If the dealer you bought the unit from won't help (he should), all you can do is to meticulously follow the instructions provided by the manufacturer and hope for the best.

Each case is different. There are no general guidelines other than "Be careful!" The way you use your Winchester will be different from the way someone else uses theirs. Your programs are likely to be different. (Some programs will not load onto a hard drive without some effort on your part; a few won't transfer at all. Others have special versions designed for transfer to a hard drive system. Ask your dealer before you buy.)

The hard drive is an important part of a system if you find that you need one. Don't take the installation lightly. Imagine losing all the programs and data on a single diskette, then multiply this by about 30 and you'll have an idea of how devastating Winchester problems can be.

MODEMS

There are two basic types of modem presently available—internal and external. Both work equally well, and each has its own advantages and disadvantages. Installation of either is as simple as plugging in a few cords.

The internal modem is generally less expensive since it doesn't require some of the circuitry of the external type. It also tucks inside the computer and out of the way. Your desk is less cluttered and there are fewer wires strung about. The internal modem also eliminates the need (and expense) for an asynchronous port card. This won't be important if you already have a serial port

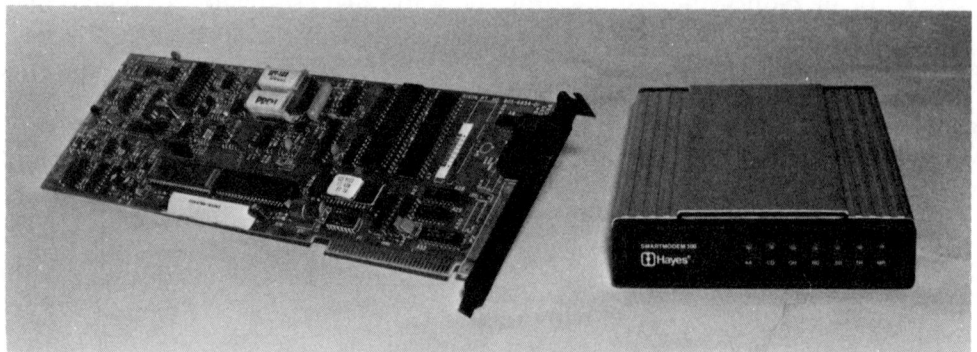

FIG. 8–16 Internal and external type modems.

Williams: Repair & Maintenance of Your IBM PC (Chilton)

ADDING TO YOUR SYSTEM

available, but if you don't have one, the expense of an external modem goes up all the more.

The major disadvantage of the internal modem is that you can't move it from computer to computer. If you have more than one computer, or if you one day decide to buy another, the modem is basically stuck where it is. A second disadvantage is that the internal type takes up one of the slots in the computer.

An external modem has the advantage of being "portable." You can move it from computer to computer by merely changing the cables from one computer to the other. The controls are on the modem, making it easier to handle and monitor.

The primary disadvantage of the external modem is that of cost. The modem itself costs more usually. If you don't already have an asychronous (serial) port, you'll have to buy one. If this is the case, you are still using up another slot in the computer. (A multifunction card can take care of this usually).

Both types of modem come in two speeds: 300 and 1200 baud, with one baud being one bit per second. For example, a 300 baud modem will transmit and receive data at 300 bits per second. The 1200 baud modem is four times faster, making it generally better. However, it's also much more expensive.

Most modems that handle 1200 baud can be switched to handle slower speeds. With them you can communicate with other computers that have only the slower models. You cannot work a 300 baud modem on 1200.

1200 baud is the usual limit for reliability across standard telephone lines. Transmission and reception of data at speeds higher than this (the computer will handle up to 9600 baud) usually results in garbled and often unusable data.

An external modem can be "direct connect" or "acoustic." The acoustic type modems have been in use for many years. They have the advantage of being very inexpensive. These modems have a cradle for the telephone. The computer generates "beeps" to represent the data, which is picked up by the telephone in the same way as it picks up your voice and other sounds.

This kind of modem has a severe disadvantage, however. As it picks up the beeps that are the data, it also picks up any other sounds in the room. This can produce garbled or unusable data on the other end. The worse the fit of the telephone receiver into the cradle, the more garbage it will pick up from the room.

Direct-connect modems eliminate this problem and are generally preferred. A line connects the modem directly to the telephone outlet in the wall. No external sounds can invade.

The most common error in modem installation is brought about by assigning the COM (communications) port to be used. The IBM PC will handle only two COM ports (COM1 and COM2). If you have two already installed and then add a modem, you could be trying to force the computer into using three communication ports, which it simply won't do. This is an especially common error when using an internal modem.

A friend of mine decided to upgrade his system. He already had a modem, operated from an asynch board. He then installed a multifunction board that had two serial ports on it. When he tried to make it work, the system would just lock up. As soon as he removed the old asynch card, everything functioned perfectly.

A similar problem can occur if you forget to tell the computer which serial port is COM1 and which is COM2. (This is done with switches, jumpers, etc. on the card.) The computer will automatically go to COM1 unless told otherwise. If you have a printer connected to this port and the modem connected to COM2, each time you try to use the modem the computer will think you're trying to do something very strange with the printer.

The DOS manual explains the COM1, COM2, and Mode. Further information will come with the modem. Learn everything you can about it before installation. If you know what your computer has installed, you'll reduce the risk of trying to get it to see a third serial port. If you happen to have more than two such ports, disable all but two. (This is usually done with a jumper on the serial board. Instructions will have come with the board.)

MULTIFUNCTION BOARDS

A multifunction board is any accessory board that provides more than one feature, such as RAM memory and a serial port on a single board. These boards have found great popularity with IBM users because of the limited number of slots inside the cabinet. The typical set-up for the PC takes up two of the five slots right away (for display monitor and drive controller cards), which leaves only three slots to handle an abundance of other possibilities. A multifunction card can take care of more than one job, and will take up only one expansion slot.

The more popular multifunction boards have RAM memory, one or more serial ports, one or more parallel ports, a clock/calendar, and possibly a game port. (Print spoolers and RAM disks are software features that come with multifunction cards that carry RAM memory. No additional hardware is needed. If you want these features, keep in mind that they "eat up" a lot of RAM. So the card you use should have enough memory to handle the jobs.)

Installing these boards can be complicated, but with a little common sense you can install one with very little difficulty.

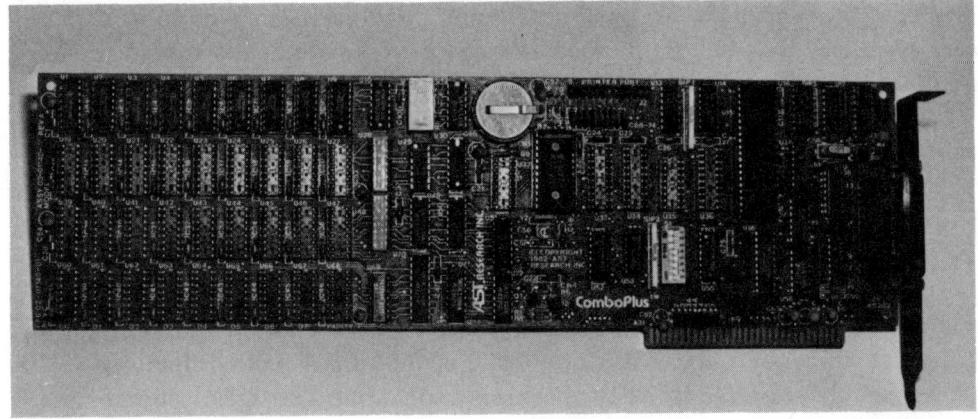

FIG. 8–17 Multifunction card. The large "can" is the battery for the clock. Be sure that the board you buy has a battery that can be changed easily. (A few have the battery soldered in place.)

As always, begin by reading through the instructions completely. Don't skim past sections just because you think you won't be needing them. You simply *can't* know too much about the board and its features.

If you're not careful when installing a new multifunction card, you can run into all sorts of problems. The two most common are conflicts in memory and hardware settings. If your system functioned perfectly before the installation, and suddenly something isn't working any more, chances are good that you've missed a switch setting or are trying to get the computer to do something it can't.

When installing RAM memory, for example, you must tell the computer the starting address of the memory on the new board. This is done by the switches on the card itself. If you have 256K on your system board, the memory on the multifunction card will have a starting address of 256K. The computer makes use of the memory in order: first that on the system board, then that on the banks of the accessory card. If you don't tell it where to shift over to the card (and correctly!), nothing will work right.

An ending address may also have to be set. This is again accomplished by switches on the accessory board. If your system board has 256K and the expansion card has an additional 192K on it, the ending address will be 448K.

The next step is to set the switches on the system board so that the computer knows how much actual memory it has to work with. If your system has available 448K as in the above example, the switches have to be set to tell the computer that 448K are available. Otherwise, it won't recognize the new memory. In other words, if the switches are set to tell the computer that there is 256K on board, it won't matter how much memory you add. The computer will still see just 256K. The setting of these system board switches is for *all* memory available, both on the system board and on the expansion card.

Many people find that accessories that worked correctly before adding to the system no longer work afterwards. If something worked before you installed a new device, and doesn't after, try putting things back the way they were. In cases like this, the chances are very good that the problem is caused by one piece of hardware interfering with another.

Hardware conflicts can cause some things that appear to be malfunctions, such as the computer refusing to work. Before you suspect the hardware, make sure that you've eliminated any conflicts between the installed devices.

A simple example is when there are more than two serial ports or more than three parallel ports. These are the maximum that the PC can support. Add more than this and the computer will be very unhappy. Further, each serial and parallel port must have its own address, such as COM1 and COM2 for the serial ports, and LPT1, LPT2, and LPT3 for the parallel ports.

Assigning the addresses is done by switches, by jumpers, or both. Going from one to the other is handled by the "Mode" command from DOS.

Things get more confusing because different boards work different ways. It may not seem that a COM port is being taken up by the board, when in fact it is. The documentation should give you all the information you need as to what the board requires as far as port assignments. Check it carefully.

It's possible that the board you choose will not be compatible with your system or with certain devices in your system. The dealer should be able to tell you what you need to know about compatibility. (The easiest way to avoid this problem is to use only known name products that have been proven to be fully compatible.)

NETWORKING

Networking computers is becoming more and more popular. This allows several computer stations to operate a centrally located device without the need for several devices, one for each station. The individual computers are all connected to a single device, which is shared by all computers attached to it.

An example might be in using a hard drive. The hard drive will contain all the programs and data needed by each of the stations. Rather than making the expensive purchase of a hard drive for each computer, the Winchester is shared by all.

Installation of a network system is not easy. The average person is almost guaranteed to have trouble. To solve this problem, buy the system *only* from a

FIG. 8—18 Networking board and cable.

dealer who offers full support from initial set-up to handling the trouble-shooting.

SOFTWARE

Although the programs you use are not actually devices, there is often a specific installation procedure to follow. The manual that comes with the software should give you all the information you need to get the program up and working.

Just as you would (or should) when installing any other accessory, read through the manual before beginning. At least go through the installation section thoroughly. Despite how it seems at times, all the information you need is there (*if* you can understand it). Some software is designed to work on a variety of computers. The manual might contain several sections, each for a different computer. If this is the case, be sure that you're reading the section for the PC.

If the program allows it, make at least one back-up copy (preferably two) before you begin. Then if you make a mistake you can go to your back-ups. Before adding new software ask yourself these questions: Do you understand the procedure for making *that* program work on *your* system, with *your* operating system? Do you have sufficient memory in both the drives and in RAM?

SUMMARY

There is very little that the PC can't do if you install the proper devices. It can be made to pretend to be a different computer by using an emulation package. It can become an artist or draftsman with the installation of a plotter.

Installation of some accessories is as easy as attaching a cable. Others require special patches to the software (either to the software that comes with the device, to the DOS, or possibly to both). A few devices require customized software to get them to go.

The dealer should be able to inform you of the difficulty in installing a particular device you want to purchase. Buy it from him and he takes on the responsibility of providing you with everything you need for proper operation. A part of the sale is to provide to the customer (free or at a reasonable price) the initial assistance needed for the installation, special software, software patching programs, and whatever else is needed.

Adding options and devices to your system is normally simple enough for the average user to handle. Getting a new monitor to work is merely a matter of plugging it in to the proper type of adapter board. A multifunction card requires that you set a few switches, but once again it is a "plug-in and go" circumstance.

A few items, such as a complex network, demand special knowledge. Even installing a hard drive can present problems if you're doing it for the first time.

Before making the purchase you should have some idea of just how difficult it is to handle the installation. Before beginning the actual installation, read through the instructions completely. Then go through the steps of installation in your head. Do you understand what to do, when, and why?

Before giving up, read through the instructions again. Most of the time the information you need is right there—*if* you take the time to dig it out.

Williams: Repair & Maintenance of Your IBM PC (Chilton)

============================

Dealing with the Technician
9

No matter how well you maintain your computer, or how much you learn about repairing it, there will be times when you will have no choice but to call in a professional (and pay those professional fees!). It can't be helped. Certain repairs demand the use of special (and expensive) equipment. Others require special knowledge that is far beyond the scope of any single book.

This book is meant to reduce to a minimum those times when you have to hire a professional, and to reduce the amount you'll have to spend when those times occur. When you have to consult a professional, you'll have already taken care of many of the diagnostic steps and can supply a great deal of information to the technician. Since you've spent the time, he doesn't have to (and you don't have to pay for the time.)

This book can go a long way toward reducing your maintenance and repair costs, but it *can't* reduce those costs to zero.

MAIL ORDER

If you make purchases through the mail the service situation is different than if you purchase through a local dealer. Very few mail order companies are set up to handle questions or problems. You get a discount price, but you are expected to take care of everything by yourself. The only responsibility that many mail order companies accept is that the equipment is "as advertised."

While they do guarantee that the equipment will arrive in good condition, it is usually assumed that the buyer knows enough about what is going on to get the equipment or software functioning.

There is nothing dishonest in this, and the lack of service isn't because the mail order companies don't care. They simply aren't set up in the same way as a local dealer. They don't have the technical staff on hand to take care of questions and problems. Their basic job is to fill your order, which they usually do quite well.

If you know for certain that you can handle any problems with the installation, and you know that you can wait for the delivery (and exchange if necessary), mail order can save you some money. If you have any doubts about your technical knowledge, you would probably be better off working with a local dealer.

RESPONSIBILITIES OF THE DEALER

When the dealer sells you some equipment or software, he assumes a certain amount of responsibility. (If he doesn't, you should probably find another dealer.) This begins with his making sure that the computer is functioning when you get it. If you buy an entire system from the dealer, he should see to it that everything functions as a unit before turning it over to you. It should *not* be handed over as nothing more than a pile of boxes. If the dealer operates this way, you might as well go through a mail order company and save some money.

The PC is quite versatile. It will accept a wide variety of options manufactured by many different companies. To attain this versatility it is sometimes necessary to make adjustments. Whether this is handled in the equipment (through DIP switches or some other way) or by the software, the dealer should take care of all the little details for you. This is actually to the dealer's benefit. He doesn't want to have his technical staff tied up on the phone telling a customer how to plug in a cable.

Many dealers include training in the purchase price. This training isn't meant to make you a computer expert. That takes years of work. The purpose of this training is to get you familiar with the computer and with the software. (Unfortunately, a lot of training simply does what the documentation *should* have done in the first place.)

Occasionally the dealer will charge a small fee for this training. This is especially true for instruction in how to operate a complicated piece of software. The more training you need, the higher the cost. These charges are normal. Even a $1000 software program doesn't automatically bring with it a complete course of instruction. At the same time, the dealer has a responsibility to show you the basics required. Preferably, when you walk out the door with the package under your arm, it should be ready to stick into the drive of your own computer without the usual hassle of initializing and installing. (Bring or buy some blank diskettes. If you wish him to install the system on the diskette, also bring along your own copy of DOS. It is technically illegal for anyone but you to put the "system" onto a diskette.)

After purchase of either hardware or software, the dealer continues to be responsible to his customers. If you have a problem a few weeks or months after getting the equipment home, you should feel free to call in with questions.

Manufacturers of both hardware and software are famous for refusing to talk to the end user. They usually assume that dealing with the end user is the responsibility of the person who sold the item. Thus, even if he doesn't want it, the responsibility falls on the dealer to provide needed information for his customers.

Don't be afraid to ask the dealer to put in writing the promises he makes for support of the products you buy from him. People who are new to computing (about 75% of the dealer's sales) won't know what questions to ask when they make the purchase. Something as simple as formatting a diskette is a major accomplishment for some newcomers. The more the buyer works with the computer, the more questions he'll have. After a short time, certain questions and problems are bound to come up.

The dealer should provide competent technical assistance after the sale. The user needs a source of information when something goes wrong. This includes technical questions involving operation of the software and hardware purchased, as well as repair when something goes wrong. (To keep the dealer on your side, try to find out if the malfunction has been caused by the operator, or by someone who hasn't bothered to read the instruction manual.)

Simple questions should be answered without charge (if you're a customer). Questions that involve some instruction or lengthy personal attention will probably involve a fee. Repairs of any size will cost unless you're under warranty or have a service contract (which also costs money).

Some smaller dealers do not keep a technical staff in the shop. They can't afford to. If this is the case, the dealer should at least be able to guide you to the proper people for your needs or to find the answers to your questions. Being small is no excuse for being unable to provide customer service when it is needed. Customer service is the responsibility of dealers of any size, and it is something you should look for when locating a dealer to do business with.

YOUR RESPONSIBILITIES

If you expect assistance or information from a local dealer, you should begin by giving him some business. If you bought your computer through the mail, don't expect the local dealer to answer all your questions and take care of your problems free of charge.

Keep in mind why the dealer is there. His goal is to earn a living, hopefully a fairly nice one. The more he gives away free, the less chance he has of surviving. The profits aren't nearly as high as you might think.

He's *not* there to provide an easy source of free advice or free services. The technical staff kept on hand to handle questions, problems, and repairs costs the dealer money. There are salaries to pay, equipment to buy and maintain, plus lots of other expenses. When the staff is doing a repair job, they are earning their keep, and the store is making money. When they're handing out free information, they're doing exactly the opposite.

Even if you didn't purchase your system from a particular dealer, you can build a working relationship by giving him your other business. The next time you're in the market for a printer, monitor, or software, stop in and see the dealer you plan to use for help. Or you might have been thinking about getting some training on how to operate one of your programs. Perhaps the dealer has a course available. However, don't expect him to be impressed because you'll be buying a $30 game every six months or a $2.95 magazine every month. (Some stores have taken out the magazine racks because the profit involved isn't worth the hassle. I hope they haven't removed the book shelves as well!)

If you have no present needs, you can still check around with the local dealers. Even something simple can get a positive response if its honest. "Everything is working just great right now, but I figure eventually I'll need someone to fix something. I'm checking around so I know who to contact when that time comes."

Keep in mind that it works both ways. Be fair with the dealer and he's more likely to be fair with you. Obviously, there is no sense in buying things you don't need. But when you *do* need something, and if the dealer has gone out of his way to be helpful, show him your appreciation by doing business with him.

THE TECHNICIAN

With the information in this book you should be able to provide a considerable amount of information to the service technician. Your goals are to reduce the cost of repair and the amount of time that repair will take. By giving the technician as much information as possible—and accurate information—you reduce both the cost and the amount of time needed to complete the repair. You're also likely to get better service by letting the technical staff know that you have some idea of what you're talking about.

Imagine yourself as the technician. You receive a call from someone who says, "My computer's broken." It's hard to get any specific information from the person because all your questions apparently make no sense to him.

A second call comes in. This time the person tells you all the things that have been done in the attempt to track down the problem. Each question you ask is answered quickly and accurately.

Of the two, which are you more likely to treat as an equal?

Having some idea of what is going on can provide better service. If the technician (and the shop) is honest and reliable just the idea that they're dealing with someone with intelligence will help. If they happen to be one of the very few dishonest shops around, the fact that you know a little something can deter them from trying to "pull a fast one."

A fair part of the technician's day is simply guiding the customer through the right questions. When the customer calls in with a "Why is it broken?" query, the technician spends some time getting the customer to answer the right questions. He knows that the customer who knows what things to look for is rare. This can be frustrating. He comes to expect a lot of "I don't know" answers from the customer. Some technicians get so frustrated that they don't even want to talk to the customers. "Just bring it in and I'll have a look."

If you've taken the time to become familiar with your system and investigate any malfunctions methodically, you can tell quite a bit about how well the technician "knows his stuff" even over the phone. His responses to your information and questions should make sense. (A response of "I can't tell without looking at it" doesn't mean much.) Of course, your information has to be accurate.

You have every right to ask about the qualifications of the person who will be working on your computer. Is the technician new to the business, with no background at all? Is his training more along the lines of television repair, or has he had schooling that concerns computers? Does he seem to understand the information you provide? Can he answer your questions without tossing around technical jargon? (On the other hand, technical "jargon" is sometimes necessary, both for you to describe the problem and for the technician to explain it.)

Don't let surface appearances fool you. A coat and tie doesn't guarantee that the technician is of high quality, nor does a tee shirt or youth mean that he can't handle the job. Also, someone with a Ph.D. in computer electronics could know less about how to get at the problem you're having than a twelve-year-old amateur. You'll learn more about the person by keeping an open mind.

Just as you have the right to inquire about the technician's background, you have the right to talk to the person who will be handling the repair or to the service manager who handles that department.

"Are you qualified?" should *not* be your conversational opener. Imagine *your* credentials being doubted right off. No matter how tactful you are, the question can seem to express criticism or doubt all too easily. Talk a little first. (Not too much. Remember, he has other things to do.) Discuss the problem. Then when you ask, "Would you mind if I asked where you got your training?" it sounds more like you're interested than worried.

Don't be afraid to make suggestions or helpful comments. It's possible that you know something special about the circumstances. If you've gathered information provide that information. Anything that makes the technician's job easier will be appreciated. It should also help to reduce the amount you pay for the technical work. This doesn't mean that you should act like a know-it-all.

Common sense and courtesy are your two greatest assets in dealing with a technician. The technician, the manager, the dealer—they're people. Treat them with respect and you'll get much more accomplished.

TERMS FOR REPAIR

Two-way communication is important in any transaction. Both parties should understand fully what is promised and what is expected before things begin. Misunderstandings can come up all too easily unless anticipated before things start.

You should have some idea of the terms and the cost before the work begins. They may not be able to tell you exactly until they've actually found the problem. (For example, a problem that you think is something minor in the disk drive might be something major on the system board.)

An experienced technician will usually be able to give you a fairly accurate estimate. This should be given to you in writing before the actual work begins. If further testing and diagnosis reveals that the problem is something more expensive, be sure to have it (in writing) that the dealer will call you before going ahead (unless you truly don't care).

Along with the price estimate, you should have a time estimate. How long will your computer be tied up? If it goes beyond this period can you get a "loaner," or at least be able to rent a machine to use while you're waiting? (Few dealers can do this, but ask anyway.)

What kind of warranty is given with the work? (It should be *at least* 30 days on both parts and labor.) Having the warranty in writing is important.

Get *everything* in writing. *Insist* on this!

When the work is complete, ask for an itemized list of what was done, and what the cost was for each thing. This is your protection for any warranty service and is also a good thing to have around for future reference. To avoid complications, it would be a good idea to make your request for the itemized list before the work begins. Some shops automatically keep an itemized list. Others do not.

Before turning over the computer, be sure that the technician knows as much as possible about the problem. Does he understand what is wrong and what the symptoms are? If not, he may not know where to begin or what to look for. The more information you can provide, the better.

Communication is important if you wish maximum efficiency and a good repair job.

SOLVING REPAIR PROBLEMS

No matter how good the technician is, or how reputable the company is, there will be times when problems will arise. Many could have been prevented if communications between you and the shop were carried on properly to begin with. Others result from unforeseen malfunctions. Or perhaps one of the parts installed is faulty. Occasionally a mistake will have been made during repair.

If the work is not done to your satisfaction, say so. But again keep in mind that you're dealing with people. The nastier you are, the less willing they'll be to take care of the problem. That's only human nature. (You'd react the same way.)

Talk to the technician who did the work first. Chances are good that the problem is something he can handle. If he can't solve the situation, go to the service manager, and then to the general manager. It might take a little longer to go through the "chain of command," but the end results are often better.

SUMMARY

Before making a purchase, determine whether you'll need the fast and expert advice that can be provided by a local dealer. You can save money by using a mail order company, but you are not likely to get much help if you run into problems with the installation. You must decide whether or not you can handle the job yourself. If you have doubts, you might be better off going with a local dealer.

Both the dealer and you have responsibilities. It's a two-way street—or should be. The dealer owes to the customers all the necessary support for whatever is being sold. He should be willing to stand behind the products carried. The staff should be competent enough to give sensible advice as to which products will best suit your needs.

While the dealer is there to serve you, he is not there to be your servant. Nor is he there to give away anything free.

The key to any relationship is to treat others as you would wish to be treated. Even the most qualified technician can make a mistake now and then. When an error is made, let the technician know about it. But you'll get a much better response if you don't let him know too loudly.

===

Troubleshooting Guide
10

This chapter is a quick guide for general troubleshooting. Sometimes it will be all you'll need. More often it will be merely the starting place. At all times you are encouraged to reread at least Chapters 1, 2, and all appropriate sections in other chapters thoroughly before getting into any repairs.

One thing has been mentioned several times throughout the book. Most malfunctions have nothing to do with the hardware but are caused by either operator or software error. Keep this in mind at all times. Completely eliminate operator and software error before you waste time on trying to repair the computer.

The inside of the computer is rather simple. As said before, it's much less complicated than the inside of your television set. There isn't as much to go wrong with it as you might think. Controlling your computer is another matter. The complexity and power of a computer rests with the person and program that operate it. If you have an operational malfunction, the chances are very good that the computer is *not* at fault.

Running the diagnostics diskette is a critical first step. Without it you are usually dealing in guesswork. (Even the fact that the diagnostics won't run tells you something.) Between this and the automatic power-on self test, you'll have most of the information you need to make the repairs.

The following table lists malfunctions by symptom, by error code, or by both. (The "x" in an error code will be a number on your screen. Don't forget to make a note of these numbers in case you have to take the computer to a professional technician.) The second column tells you the probable cause(s). The third column gives you a brief account of what you can do about the problem, plus the section(s) of the book to which you can turn for further details.

Quick-Reference Table for Locating Source of Malfunctions

Symptom	Cause	Cure
Nothing happens —or— Continuous beep —or— Series of short beeps —or— "Parity Check x" —or— 02x	Power	Check plug Check outlet Check switch Check power supply See Ch. 1 ("Checking for Power"); Ch. 4 ("Pin Voltages"); Ch. 6 ("Power Supplies")
1 long beep, 1 short —or— 1xx (101, 131) —or— Improper read/write	System Board	Check connectors See Ch. 5
Improper read/write —or— 20x (201) —or— xxxx —or— xx20x —or— xxxx201 and Parity Check x	Memory	Check switches Check connectors See Ch. 5
Improper display —or— Function error —or— 30x (301) —or— xx30x (xx301)	Keyboard	Check connector See Ch. 6 ("Keyboards")
1 long beep, 2 short —or— 1 short beep and incorrect or no display	Display	Check connectors Check power (Ch. 6) Check switches
4xx	Monochrome	
5xx	Color	

Williams: Repair & Maintenance of Your IBM PC (Chilton)

TROUBLESHOOTING GUIDE

Quick-Reference Table for Locating Source of Malfunctions, *continued*

Symptom	Cause	Cure
BASIC on display —or— Improper read/write —or— 6xx (601)	Drive	Check Doors Check software See Ch. 4
7xx	Math coprocessor (8087)	See Ch. 6 and 8
Printer error —or— 9xx	Printer or adapter card	Check connectors Check printer See Ch. 6 and 8
11xx —or— 12xx	Asynch communications	Check connections Check adapter cards Check devices
13xx	Game Adapter	Check connections Check joystick Check card
15xx	SDLC comm. adapter	Check connections Check devices Check software Check card
Improper function —or— 17xx (1701)	Hard Disk	Check connections Check software Check drive See Ch. 8
18xx (1801)	Expansion unit	Check connectors Check devices installed See Ch. 8
20xx 21xx	Bisynch adapter	Check connections Check devices Check card

==

Helpful Charts and Tables

Self Test Error Codes

Note: An "x" indicates any number. Make a note of the actual error code that shows.

Code or Signal	Meaning	Code or Signal	Meaning
No display, no beep	Power	2xx	Memory
Continuous beep	Power	3xx or xx3xx	Keyboard
Repeating beeps	Power	6xx	Drive
1 long, 1 short	System board	xxxx201 & "Parity Check x"	Memory
1 long, 2 short	Monitor	Parity Check x	Power
1 short, wrong display	Monitor	17xx	Hard drive
1 short, BASIC statement	Drive	18xx	Expansion unit
1xx	System board		

Diagnostics Error Codes

Note: An "x" indicates any number. Make a note of the actual error code. If the final two digits are zeroes, that device has tested successfully.

Code	Meaning	Code	Meaning
02x	Power	11xx	Asychronous adapter
1xx	System board	12xx	Alt. asynch adapter
20x or xxxx201	Memory	13xx	Game adapter
30x or xx30x	Keyboard	14xx	Printer
4xx	Monitor (monochrome)	15xx	SDLC communications adapter
5xx	Monitor (color)	17xx	Hard drive
6xx	Drive	18xx	Expansion unit
7xx	Math coprocessor	20xx	BSC adapter
9xx	Printer adapter	21xx	Alt. BSC adapter

Williams: Repair & Maintenance of Your IBM PC (Chilton)

Disk Drive Error Codes

Code	Meaning
606, 611, 622, 623, 624, 625, 626	Signal cable, adapter, or drive assembly
607	Write-protect switch
608	Diagnostic-disk failure
612, 613	Cable or adapter

Power to Drives

Com (−)	Probe (+)	Voltage dc
2	4	4.8–5.2
3	1	11.5–12.6

Note: These voltages and pin numbers are the same for testing power to a hard drive.

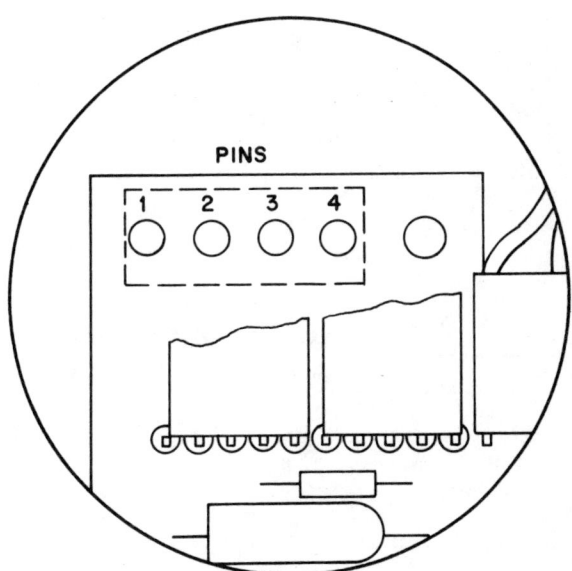

FIG. A–1 Pin locations for checking power to drives.

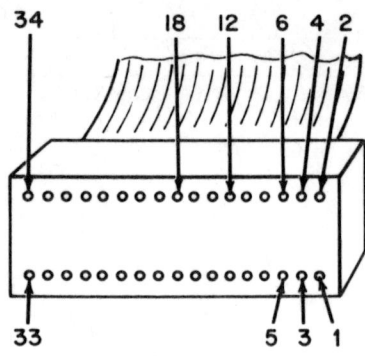

CABLE CONNECTOR
PIN LOCATIONS

FIG. A–2 Signal cable pin locations.

Signal Cable Pins—Twisted Pairs to Drive A

Note: The cable to drive A has a few twisted wires. To test continuity, all pins correspond (i.e., pin 1 on one side of the cable matches with pin 1 on the other) except pins 10–12 and pins 14–16.

Pin	Connected to Pin	Pin	Connected to Pin	Pin	Connected to Pin	Pin	Connected to Pin
1	1	2	2	19	19	20	20
3	3	4	4	21	21	22	22
5	5	6	6	23	23	24	24
7	7	8	8	25	25	26	26
9	9	10	16	27	27	28	28
11	15	12	14	29	29	30	30
13	13	14	12	31	31	32	32
15	11	16	10	33	33	34	34
17	17	18	18				

Voltages at System Board Power Connector

Com (−)	Probe (+)	Voltage (dc)
4	8	10.8–12.9
5	1	2.4–5.2
5	10	4.8–5.2
7	3	11.5–12.6
9	6	4.5–5.4

Williams: Repair & Maintenance of Your IBM PC (Chilton)

HELPFUL CHARTS AND TABLES

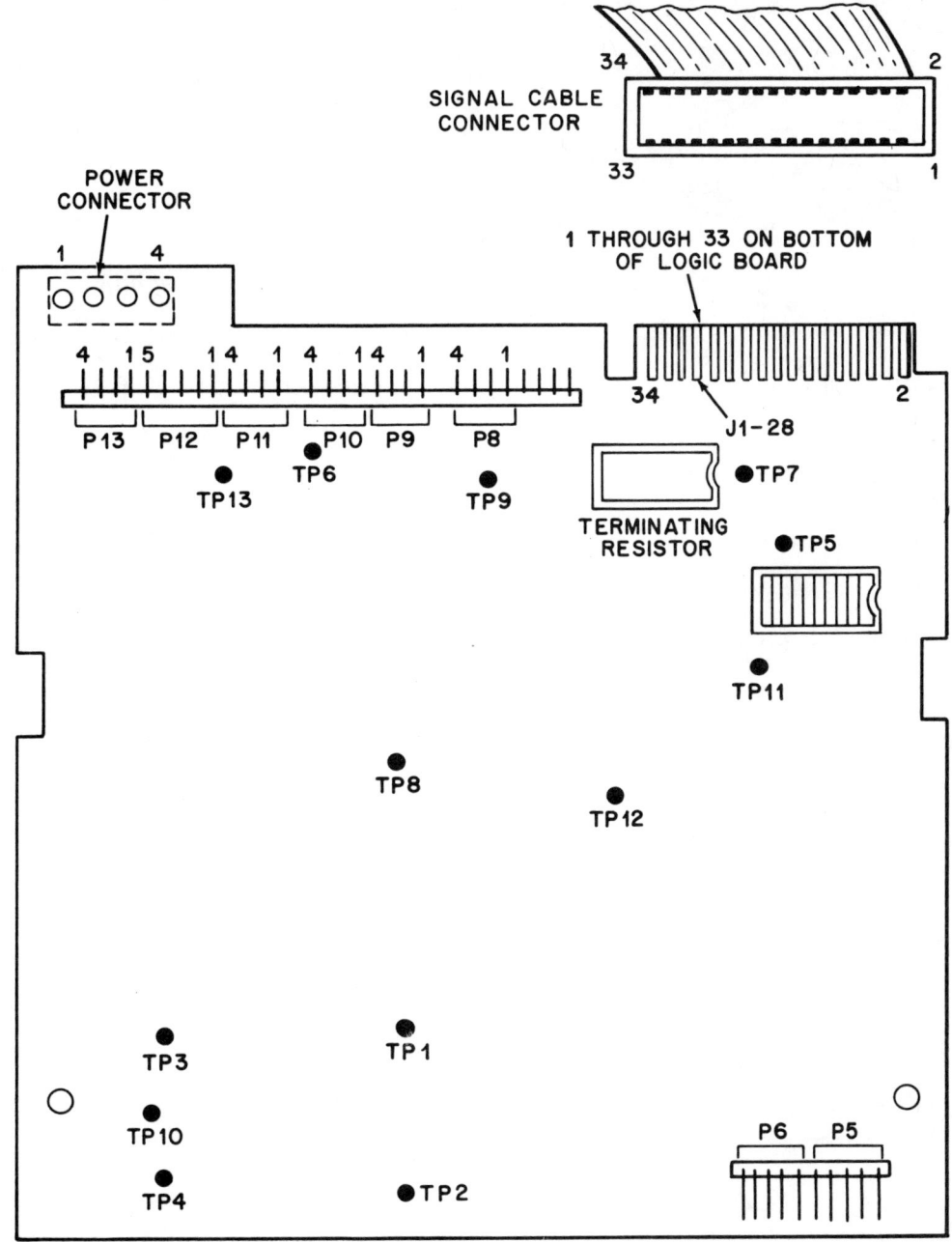

FIG. A–3 Tandon (Type 1) drive with pin and test point locations.

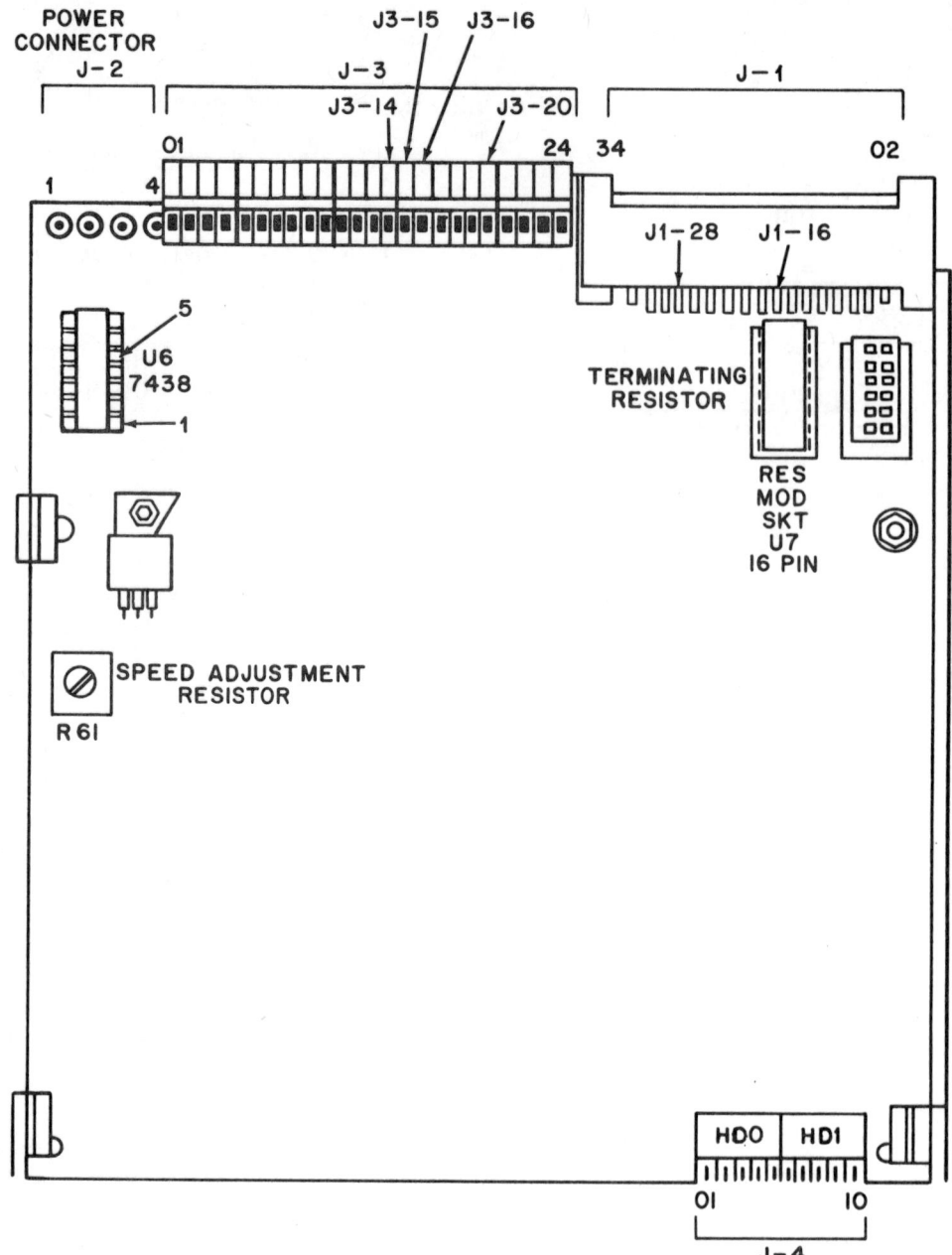

FIG. A–4 CDC (Type 2) drive with pin and test point locations.

Williams: Repair & Maintenance of Your IBM PC (Chilton)

HELPFUL CHARTS AND TABLES

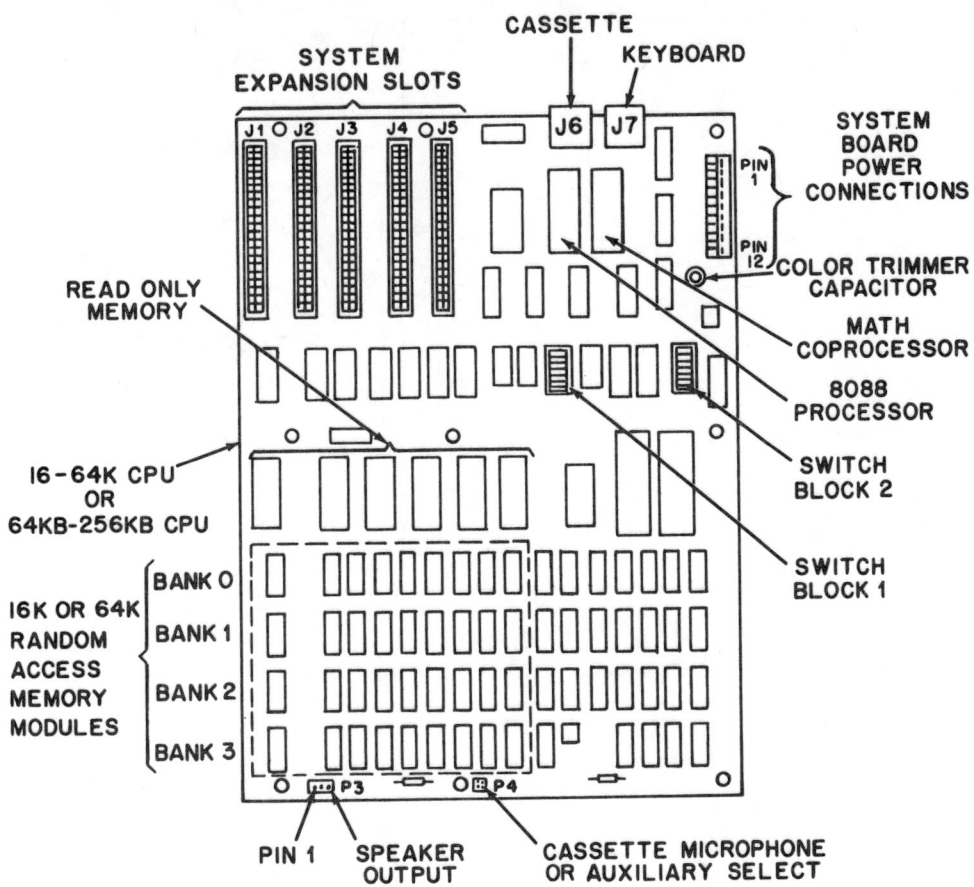

FIG. A–5 PC system board.

Resistance at System Board Power Connector

Note: Power must be off and the cable disconnected.

Com (−)	Probe (+)	Minimum Ohms
5	3	6
6	4	48
7	9	17
8	10	0.8
8	11	0.8
8	12	0.8

Williams: Repair & Maintenance of Your IBM PC (Chilton)

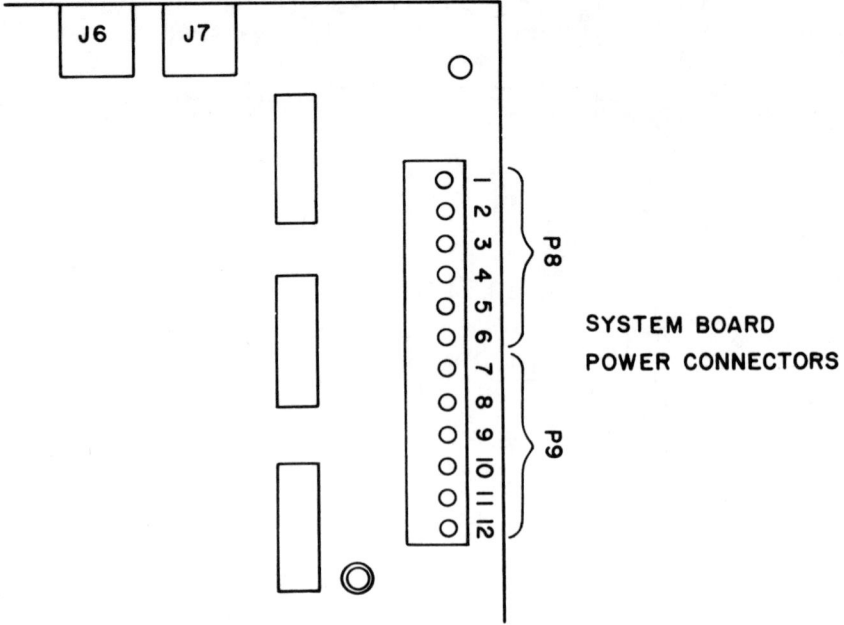

FIG. A–6   Pin locations on system board power connectors.

External Drive Port Outputs

Pin	Signal	Pin	Signal
1–5	Unused	13	Select head
6	Index	14	Write enable
7	Motor enable C	15	Track O
8	Drive select D	16	Write-protect
9	Drive select C	17	Read
10	Motor enable D	18	Write
11	Stepper motor	19	Unused
12	Step pulse	20–37	Ground

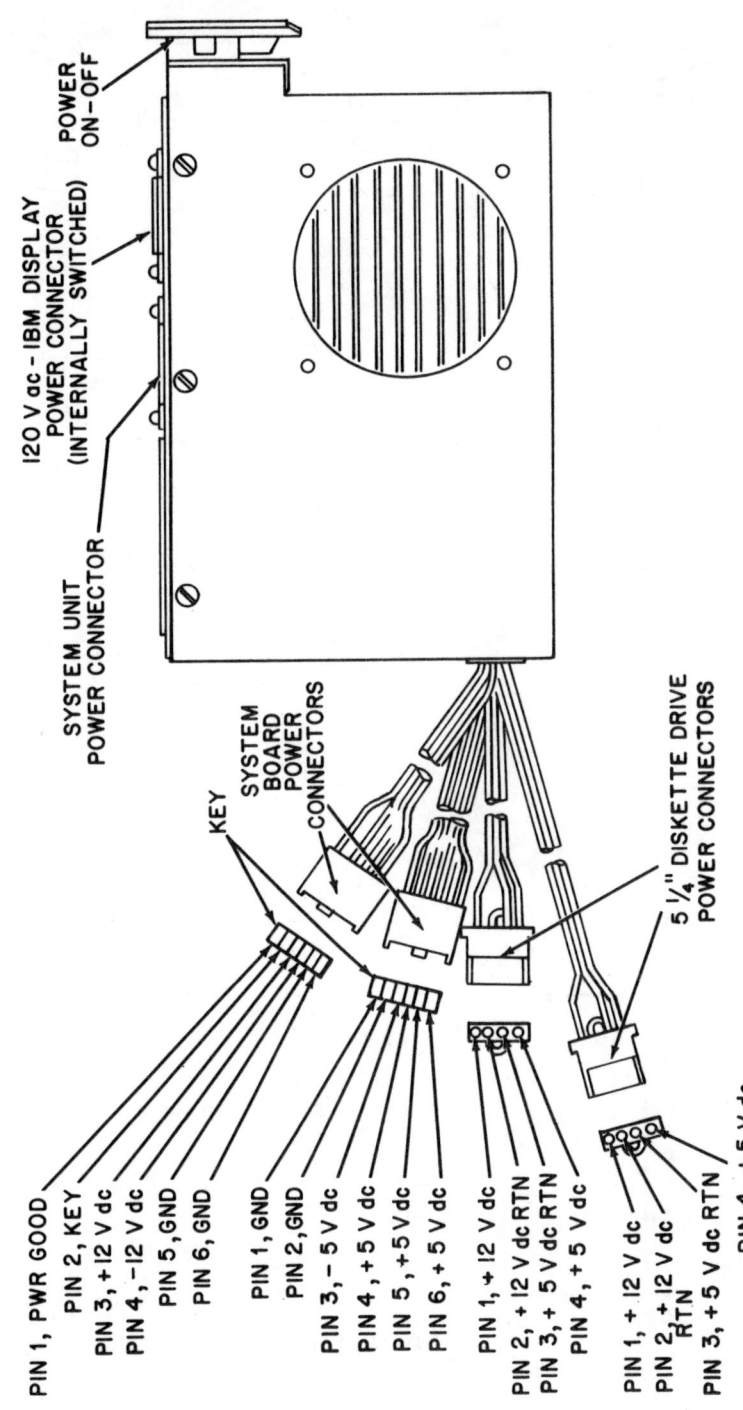

POWER ON-OFF

120 V ac - IBM DISPLAY POWER CONNECTOR (INTERNALLY SWITCHED)

SYSTEM UNIT POWER CONNECTOR

KEY

SYSTEM BOARD POWER CONNECTORS

5 1/4" DISKETTE DRIVE POWER CONNECTORS

PIN 1, PWR GOOD
PIN 2, KEY
PIN 3, +12 V dc
PIN 4, -12 V dc
PIN 5, GND
PIN 6, GND

PIN 1, GND
PIN 2, GND
PIN 3, -5 V dc
PIN 4, +5 V dc
PIN 5, +5 V dc
PIN 6, +5 V dc

PIN 1, +12 V dc
PIN 2, +12 V dc RTN
PIN 3, +5 V dc RTN
PIN 4, +5 V dc

PIN 1, +12 V dc
PIN 2, +12 V dc RTN
PIN 3, +5 V dc RTN
PIN 4, +5 V dc

FIG. A-7 Power supply and outputs.

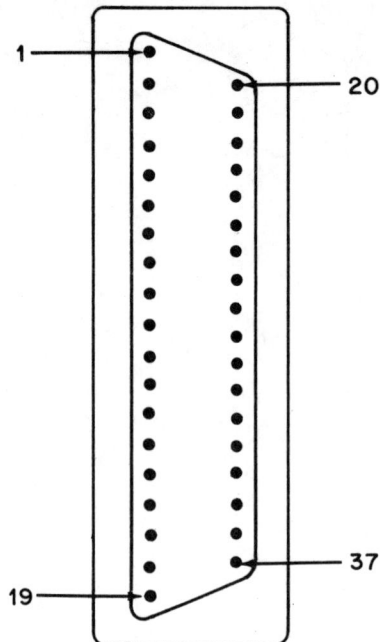

FIG. A–8 External drive port pin locations.

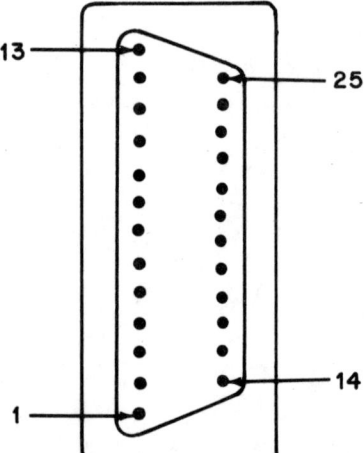

FIG. A–9 Serial communications drive pin locations.

Serial Port (Asynch) Outputs

Pin	Signal	Pin	Signal
1	Unused	11	−Transmit current loop (send)
2	Transmit	12–17	Unused
3	Receive	18	+Receive current loop (send)
4	Request to send	19	Unused
5	Clear to send	20	Data terminal ready
6	Data set ready	21	Unused
7	Signal ground	22	Ring indicate
8	Carrier detect	23–24	Unused
9	+Transmit current loop (return)	25	−Receive current loop (return)
10	Unused		

Parallel Printer Adapter Output

Pin	Signal	Pin	Signal
1	−Strobe	10	−Acknowledge
2	+Data bit 0	11	+Busy
3	+Data bit 1	12	+Out of paper
4	+Data bit 2	13	+Select
5	+Data bit 3	14	−Auto feed
6	+Data bit 4	15	−Error
7	+Data bit 5	16	−Initialize
8	+Data bit 6	17	−Select
9	+Data bit 7	18–25	Ground

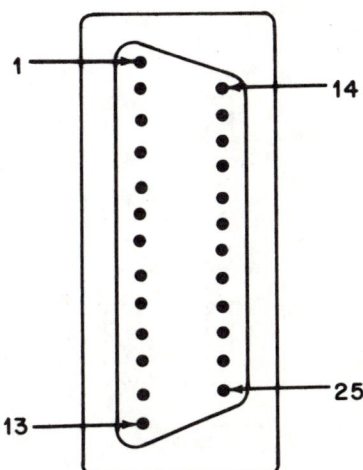

FIG. A–10 Parallel printer pin locations.

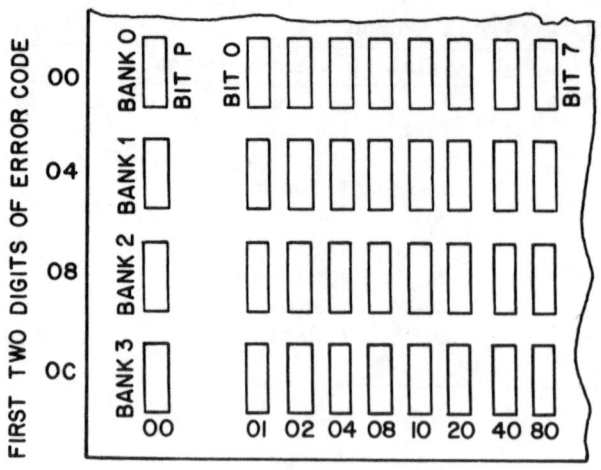

FIG. A–11 16KB–64KB system board RAM addresses.

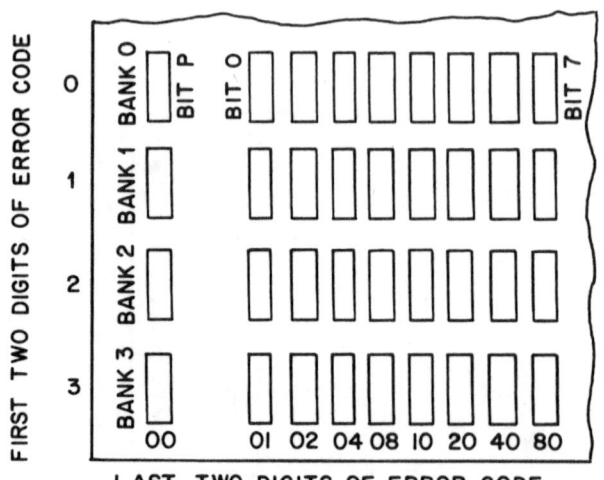

FIG. A–12 64KB–256 KB system board RAM addresses.

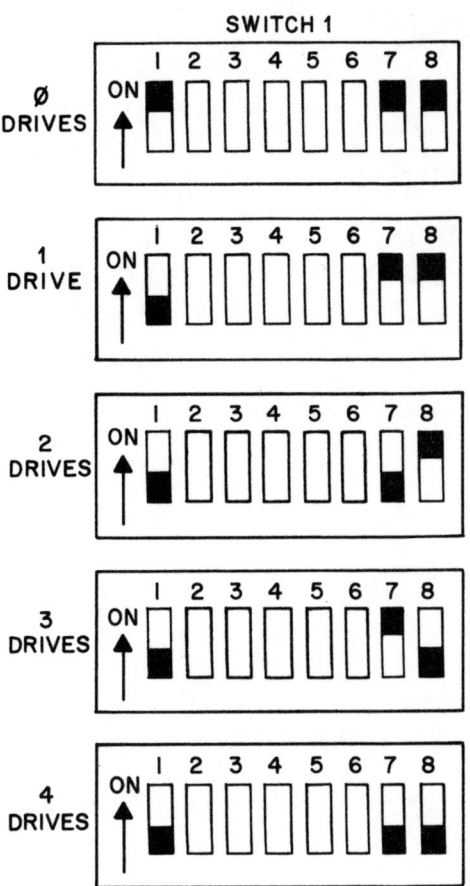

FIG. A–13 5¼-inch diskette drive switch settings.

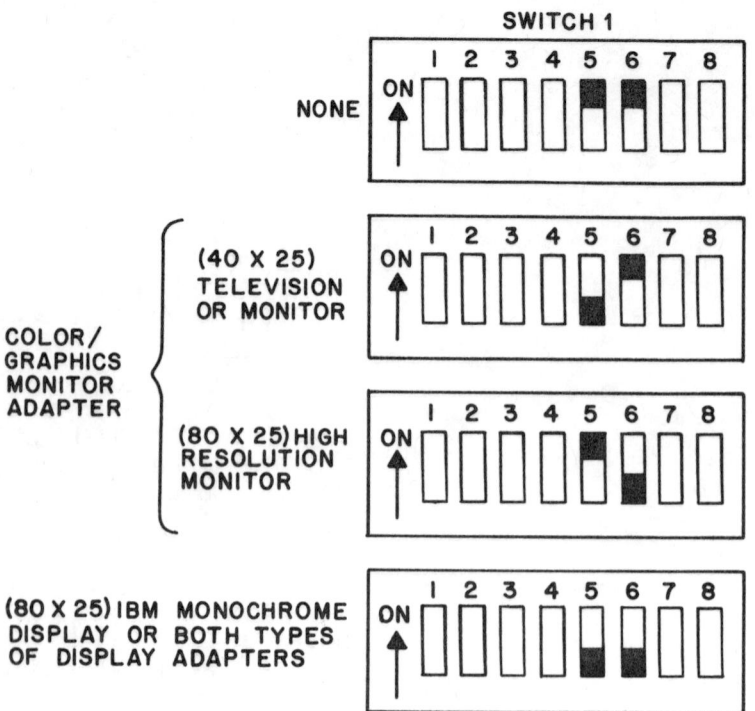

FIG. A–14 Monitor type switch settings.

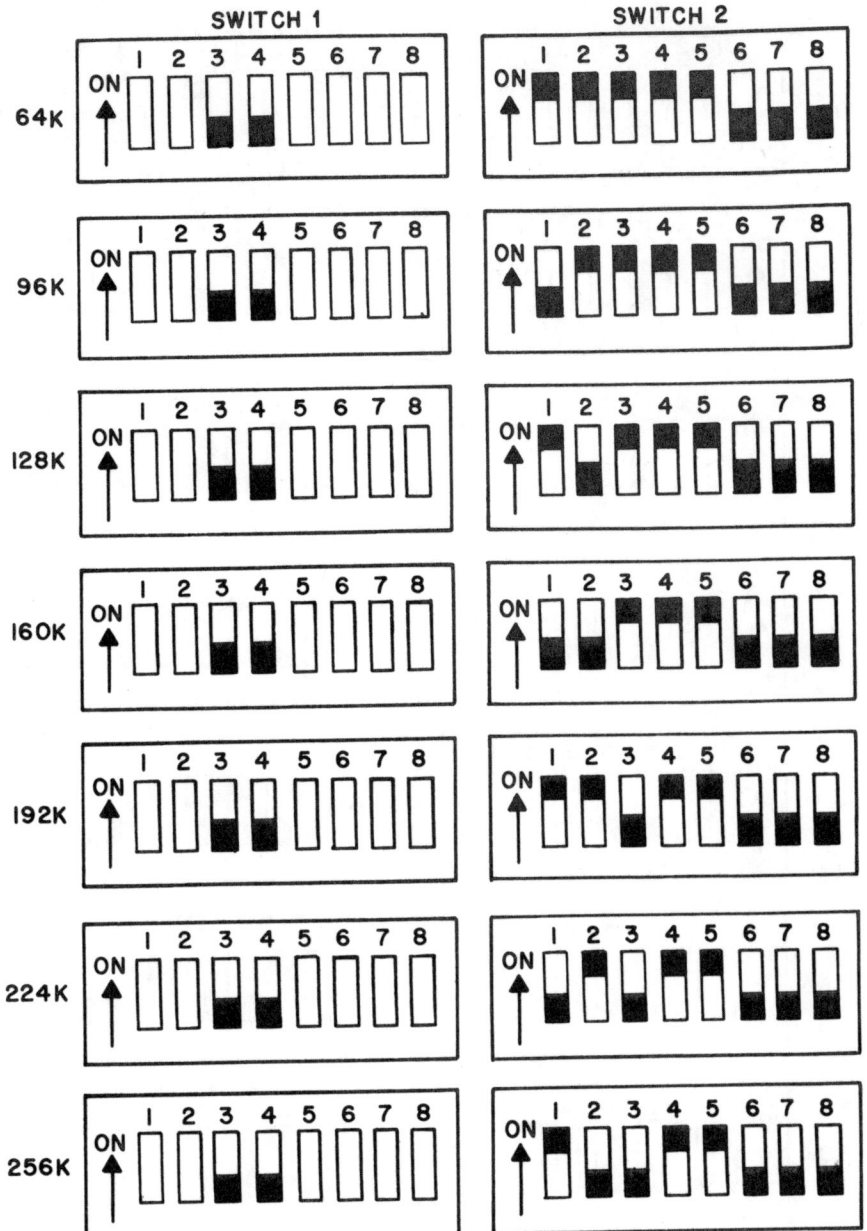

FIG. A–15 System board memory switch settings.

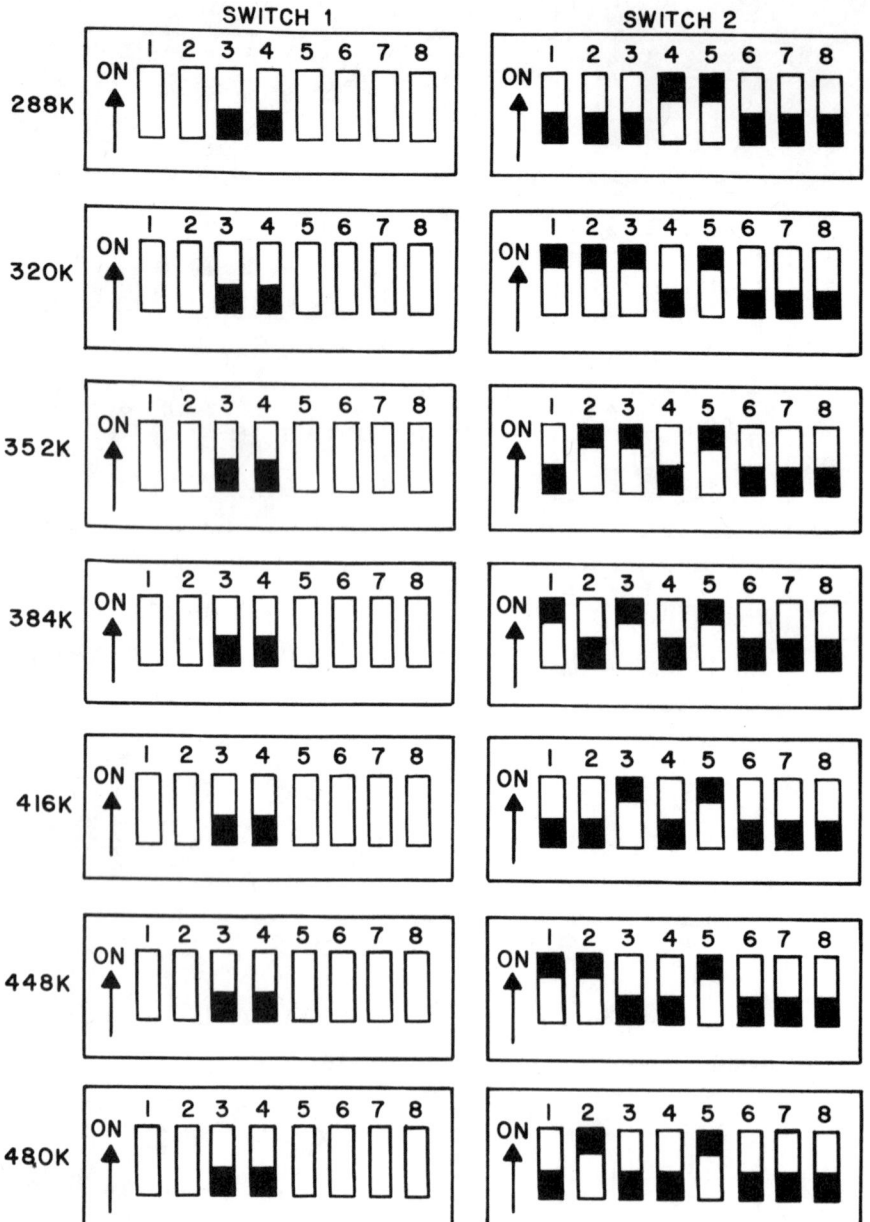

FIG. A–15, continued. System board memory switch settings.

Williams: Repair & Maintenance of Your IBM PC (Chilton)

HELPFUL CHARTS AND TABLES

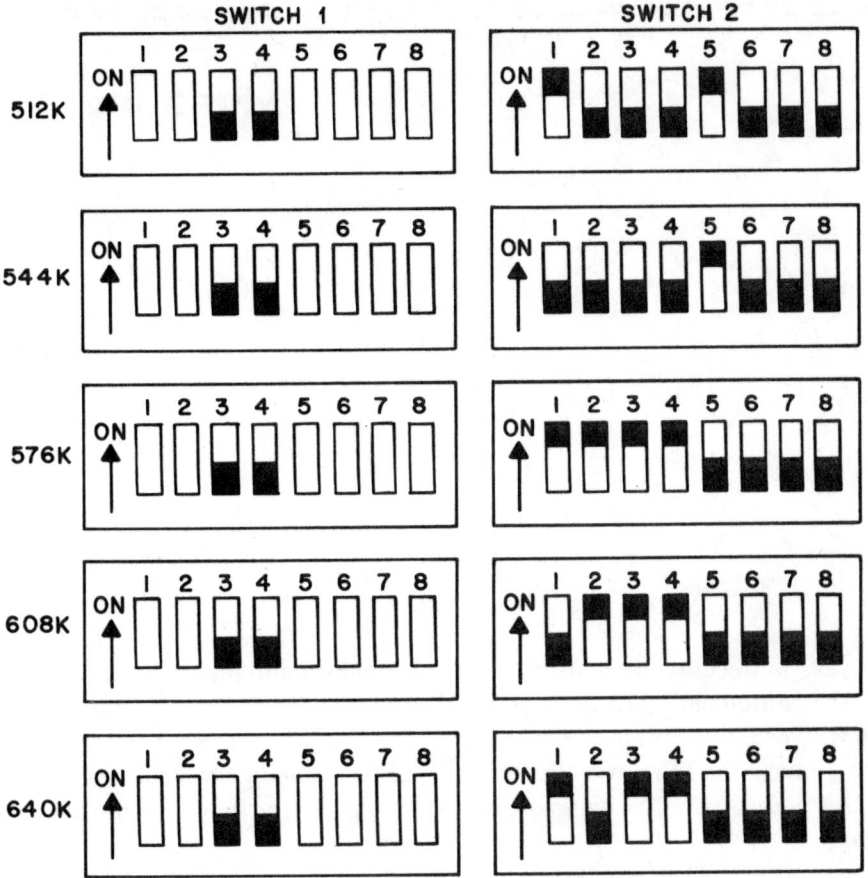

FIG. A–15, continued. System board memory switch settings.

Glossary

AC: Alternating current measured in cycles per second (cps) or hertz. The standard value coming through the wall outlet is 120 volts at 60 hertz, through a fuse or circuit breaker that can handle about 15 amps (check for yourself to know). The PC can tolerate an ac value between 104 volts and 127 volts. The power supply changes this to the proper dc levels required by the computer.

ACCESS: A fancy term for "to get at."

ACOUSTIC: Having to do with sound waves. For example, an acoustic modem (the kind that resembles a telephone) sends and receives data as a series of audible beeps. (A direct-connected modem is better than an acoustic modem since it is not prone to interference or false signals due to room noise.)

ADDRESS: Where a particular piece of data or other information is found in the computer. See Chapter 5. Can also refer to the location of a set of instructions.

ASCII: Acronym for American Standard Code for Information Interchange. This code assigns binary (on/off) values to the seven-bit capability of the computer (plus the extra 8th bit to signal the end of the character or to be used as a parity bit to check for errors). ASCII is the standard code used to send data and other binary information, such as through a telephone modem.

ASYNCH: An abbreviation for "asynchronous," generally applied to communications and the way in which a character is transmitted and checked. Each character is "balanced" individually, such as with a stop bit.

AUDIO: A signal that can be heard, such as through the speaker of the PC. Diagnostics uses both visual codes (on the screen) and audio signals.

AUTOEXEC: A special batch file generally used to start a program automatically when inserted in the drive. See your DOS manual for how to create such a file.

BACK-UP: A copy of a program or data diskette. Make them often to protect yourself.

BANK: The collection of memory modules that make up a block of 64K in RAM, usually nine ICs.

BASIC: One of the most common computer languages learned. The IBM PC has three versions: cassette, disk, and advanced. Cassette BASIC resides in ROM and is loaded automatically if drive A is empty (or thinks that it is empty). Disk BASIC and advanced BASIC (BASICA) are on the DOS diskette.

BAUD: A measure of the rate of data transmission. The transmitting signal is split into a certain number of parts (usually bits) per second. A rate of 300 baud means that 300 bits per second are being sent.

BISYNCHRONOUS: A method of communication between a mainframe computer and a minicomputer. (See also "asychronous.")

BIT: The single pulse (on/off) of information used in binary code. The word "bit" is actually an abbreviation for "binary digit."

BOOT: To load a program into the computer. The term comes from "bootstrap," which in turn comes from "lifting oneself by one's own bootstraps." In simple terms, it means that the computer is loading itself and is setting the computer to operate, without other operator intervention.

BUFFER: A segment of memory used to store data temporarily while the data is being transferred from one device to another. A common example is a printer buffer. This device stores the incoming data at full computer speed and sends it to the printer at a speed the printer can use (such as 40 cps, or about 300 baud).

BUG: An error in a program.

BYTE: A collection of bits that makes up a character or other designation. Generally a byte is eight data bits, plus one parity (error checking) bit; the binary representation of a character.

CARRIER: The reference signal used for the transmission or reception of data. The most common use of this signal with computers involves modem communications over phone lines. The modem monitors this signal to tell if the data is coming through. Generally, if the carrier isn't getting through, neither is the data.

CHIP: Another name for an IC, or integrated circuit.

CIRCUIT: A complete electronic path.

CIRCUIT BOARD: The collection of circuits gathered together on a sheet of phrenolic plastic, usually with all contacts made through a strip of pins. The circuit board is usually made by chemically etching metal-coated phrenolic plastic.

COMMON: The ground or return path used to make measurements with the multimeter. The black probe.

CONTINUITY: In electronics, an unbroken pathway. Testing for continuity normally means testing to find out if a wire or other conductor is complete and unbroken (by measuring 0 ohms). A broken wire will show infinite ohms.

CPS: Cycles per second. See "hertz."

CRT: Cathode ray tube; basically a fancy name for a television or monitor screen tube.

CYLINDER: A pair of tracks on a diskette (one on each side) that are the same distance from the center of the diskette. A double-sided diskette has 80 tracks, but 40 cylinders. Cylinders are used to reduce the amount of movement of the read/write heads. The sectors of a track on side 0 are filled with data first. Then the same track, but on the opposite side, is filled, without any movement of the heads. This is why it is impossible for a single-sided drive to read a diskette that was recorded on a double-sided drive.

DATA: Information.

DC: Direct current, such as that provided by the power supply. (Also found in batteries.)

DEBUG: To rid a program of errors. Also used as the name of a program on the PC-DOS diskette that does this task.

DEFAULT: An assumption the computer makes when no other parameters are specified. For example, when you type "dir" without specifying the drive to search, the computer automatically goes to the "default" drive (normally drive A) and assumes that this is what you want. The term is used in software to describe any action the computer or program takes on its own with imbedded values.

DENSITY: The amount of data that can be packed into a given area on a diskette. Diskettes used for the IBM PC are "double density."

DIRECTORY: The allocation track of a diskette. It stores the titles given to the files saved on the diskette and tells the computer and drives how to get to those files. The directory serves as a "table of contents" for the files saved on the diskette. (Sometimes VTOC, or volume table of contents, is used instead of the word "directory.") The directory sorts data that identifies the files by name, by size, by the kind of file stored (i.e., system file, hidden file, user file), the date the file was created, and the data the computer needs to find that file on the diskette.

DISKETTE: The flat magnetically coated media used most often in computers for the storage of data.

DOS: Disk Operating System. Most common for the PC are PC-DOS versions 1.0, 1.1, 2.0, and 2.1, and CP/M-86.

DOUBLE-SIDED DRIVE: A disk drive with read/write heads both above and below the diskette. Thus, both sides of the diskette can be used.

DRIVE: The device used to read and write on diskettes. With the PC the primary drive is A:, with the secondary drive being B:. The drive may also be a "fixed" drive, more commonly called a "hard drive" or "Winchester drive." The drive can also be electronic, using available RAM instead of the magnetic media.

DSDD: Double sided, double density. Describes the diskettes used in a PC double-sided drive.

EBDIC: Extended Binary Coded Decimal Information Code. A standard method of encoding in IBM computers, but not used in the IBM PC.

EMULATE: A fancy word for "pretend to be." Often used to describe a device that is designed to make the computer *seem* to be another computer or terminal.

EDIT: To make a change or modification in data.

EXECUTE: To start a program or instruction set.

FAT: Files Allocation Table. An area on the diskette used to allocate space for files. The information included in the table is whether the diskette is single- or double-sided, eight or nine sectors per track, if the media is a fixed (hard) disk or floppy, and other information that tells the computer of files already in existence and of sectors on the diskette that are damaged or are otherwise unusable.

FCC: Federal Communications Commission. Regulates the kind and amount of radio frequencies that can be emitted by computers and computer devices.

FERRIC OXIDE: The iron substance most often used as the magnetic medium on diskettes. Abbreviated FeO. Essentially it is nothing more than rusted iron.

FIRMWARE: The ROM of the computer.

FILE: Any collection of information saved on a diskette. The file can be data, a program to run, or both.

FLIPPY: A diskette with notches and index holes cut into both sides, allowing the diskette to be used on both sides by a single-sided drive (by flipping the diskette over). Also called "flippy floppy."

FONT: A style of lettering. Many matrix printers are capable of using a variety of print styles, changeable through software.

FORMAT: A program on the DOS disk that assigns various tracks and sectors to new diskettes. Also, a particular manner in which something is laid out, such as in a program.

GROUND: The common or return side of a circuit. See "common."

HARDWARE: The computer and computer devices.

HEAD: The read/write head of the disk drive.

HERTZ (HZ): A measure of frequency. One cycle per second (cps). Frequency is also measured in units such as kilohertz (KHz, thousands of cycles per second) and megahertz (MHz, millions of cycles per second).

HUB RING: A plastic reinforcing ring applied to the spindle access hole of a diskette. This ring adds strength and improves centering.

KEYBOARD: Primary means of manual input to a computer.

IC: Integrated circuit. This is a package of electronics, often encased in black plastic, with pins coming from the bottom. Pin 1 is the first pin on the left on the side with the notch or other marking.

INDEX HOLE: Allows the computer to find the beginning of a sector on a diskette. A soft sector diskette has one physical index hole (two on a flippy floppy). Other index "holes" on the diskette are placed magnetically by the "Format" program.

INTERFACE: A fancy term for "connect." Used to describe any connection from hardware to hardware, from hardware to software, or even from hardware and/or software to the user.

K: With computers, usually used to describe an amount of memory. Normally "K" denotes 1000 (such as in kilohertz). Due to the electronics involved, 1K of memory is actually 1024 bytes; 64K is actually 65,536 bytes.

LED: Light emitting diode, such as those used in the drives to indicate that the spindle is turning.

LIGHT PEN: An instrument designed to read areas on the screen by pointing at them. Used to input data to the computer in addition to or instead of a keyboard. See also "mouse."

MATRIX: In computer printers a pattern of dots used to make up letters, numbers, and other symbols.

MINIFLOPPY: Still another name for a diskette. Describes a 5¼-inch floppy.

MODEM: A device for transmitting data over telephone lines. The name means "modulate-demodulate."

MONITOR: A television-like device to display characters and other symbols. Also called a display, CRT, or VDU.

MOUSE: A hand-held device used to input data to the computer by rolling across a surface, such as the surface of the monitor. Often used to get quickly from one spot on the screen to another. The name is derived from the small size of the device, the tail-like cable, and the two buttons that resemble (to some people) eyes.

MULTIMETER: A testing device to measure volts and ohms across a variety of ranges. Often called a VOM.

OHM: A measure of resistance.

OPERATING SYSTEM: DOS, for example.

PARALLEL: A means of data transfer with the information being sent a byte at a time. See also "serial."

PARITY: A means of error checking. Parity means "equality." Parity checking means that individual data bits are checked after a period of time to see if the bits have changed or if they are equal to what they were at input.

PC: Personal Computer.

P.E.T.:   Polyethylene terephthalate, the generic name of the plastic used to make diskettes. Common trademark name is Mylar (owned by Du Pont).

PLATEN: The rubber roller of the printer.

PORT: A place where cables are connected to the computer or other device. Sometimes called an interface.

PROGRAM: A set of instructions the computer can understand and act upon to perform some task.

PROMPT: A symbol on the screen indicating a state of readiness in the computer. A prompt indicates the computer is waiting for you to do something.

SECTOR: An area on the track of the diskette assigned to hold a certain amount of information. In the PC, each sector is assigned to hold 512 bytes of information. DOS 1.1 assigns eight sectors per track; DOS 2.0 assigns nine sectors per track.

SERIAL: A means of data transfer in which information is handled a bit (or pulse) at a time, with each bit following the others.

SOFT SECTOR: A method of setting up a diskette (through the "Format" program) so that data is written first to a sector whose position is determined by a physical index hole and thereafter to sectors marked magnetically (by "Format") successively across the track. This allows the computer to assign the sector size required. A hard sector diskette has a series of physical index holes and is usable only on specific machines.

SOFTWARE: Computer programs, usually on diskette.

SOURCE: Where a signal originates. For example, if you are copying a program, the drive containing the original is the *source*; the drive containing the diskette on which a copy is to be made is the *target*.

SPINDLE: The device in the disk drive that is inserted into the center hole and causes the diskette to spin.

SSDD: Single sided, double density. Describes diskettes used in a single-sided drive.

STEPPER MOTOR: Used to move the read/write heads across the surface of the diskette.

SYSTEM BOARD: The main circuit board of the PC. Sometimes and inaccurately called the "mother board."

TARGET: Where a signal terminates. For example, the drive or diskette on which a copy is to be made. See *Source*, above.

TERMINAL: In a computer system, a device through which data can be entered or retrieved. Can also be the end point of an electrical connection. Sometimes also used to describe a monitor (e.g., "video terminal").

TERMINATING RESISTOR: A resistor package located on the disk drive. It looks like an IC. Its function is to tell the computer which of the drives is A and B.

TPI: Tracks per inch. Standard with the PC is 48 TPI, with 40 of the tracks being used.

TRACK: One of the concentric rings on a diskette. The PC uses 40 tracks, each holding either eight or nine sectors depending on the DOS being used. Each sector holds 512 bytes of information. See also "sector."

TVI: Television interference. Computers and some computer devices emit radio frequencies that can interfere with normal television operation.

VOM: Volt-ohm-milliammeter. Commonly called a multimeter.

WRITE-PROTECT: A way to prevent recording on a diskette. The notch cut into the diskette allows a switch in the drive to activate the recording head. With this notch covered (e.g., with tape), the recording head is disabled, making recording onto the diskette impossible (hopefully).

Index

For definitions of terms, see Glossary, page 196.

cassette, as hard disk back-up, 161
Cassette BASIC, 59, 62
cassette port, 59, 100
cathode ray tube (CRT), 11
chips. *See* Integrated circuits
Chkdsk command, 47, 54
circuit boards, 95–110
    on disk drive, 73, 75, 82
cleaning
    of drive heads, 85–86, 141–143
    of expansion board contacts, 109, 135, 136
    of printer, 132, 140
cold, effect on diskette, 54
communications port, installation errors due
    to, 163–164
components
    cost of replacing, 21
    effect of current while removing, 20
    installation procedures, 148–168
    replacement, 20–21
computer safety, 15–20
computers, enemies of, 26
connectors
    as cause of malfunction, 31
    for computer components, 16
    notes concerning, 33
Control Data Corporation drives. *See* CDC
    drives
Copy command, 143
    on diagnostics disk, 35
    for hard drive back-up, 160–161
    to prevent overediting, 57
copy protected programs, 143
COPYII PC, 143
Ctrl-Alt-Del reset, 95
current
    effect on expansion board during removal,
      20
    effects of, 9
    potential for, 13
customer service, from dealer, 171

DIP switches. *See* Switches
DOS diskette, as repair tool, 4
daisywheel, cleaning, 132
danger spots, in computer, 9–11
data errors, from disjointed files, 57, 143
dealer
    support for software, 55–56, 162
    support from, 170–171
desoldering tool, 7

diagnostic diskette, 24–25, 34–52
    advanced, 84, 144
    and booting problems, 56
    for power supply test, 116
    as repair tool, 4
    screen displays, 37–38
diagnostics, 28–43
    for blank monitor, 136–137
    for disk drives, 64–84
    error codes, 36, 180
    false readings from options, 111
    for memory, 104–107
    for power supply, 112–123
    as preventive maintenance, 144
    for printers, 132–133
digital circuits, soldering iron for, 22
digital soldering tool, 5
direct-connect modems, 163
directory track, and diskette life expectancy,
    143
dirty power, 32
disjointed files, and data errors, 143
disk drive analyzer program, 84, 85, 86, 144
disk drive heads, 46, 82
    cleaning, 141–143
    cleaning kit, 85–86
    dirt on, 26, 50, 62
    noise from, 47
    not checked in self-test, 33
    program to test for alignment, 85
disk drives, 59–94
    adaptor card for, 75, 82, 157
    broken door replacement, 62, 63
    checked in self-test, 33
    circuit board, 73, 75, 82
    circuit board modification for drive B, 150
    copying diskettes prior to realignment, 86
    damage during removal and installation, 16
    diagnostics for, 64–84
    disassembly of, 87, 90–94
    display if malfunction in, 34
    error codes, 181
    half-height, 150
    installing second, 149–157
    malfunctions in, 61–64
    noise from, 47
    pin locations for checking power to, 181
    power, 120–121, 181
    power check, 69, 71–73
    power connection to, 121
    preventing problems with, 85–86

Williams: Repair & Maintenance of Your IBM PC (Chilton)

half-height disk drive, 150
hard drives, 159–162
    installation process, 161–162
    self-test check of, 33
    tape back-up for, 160
hard sector diskette, 47
heads. *See* Disk drive heads
heat, effect of diskette, 54
Heath Company, soldering course, 21
hertz, 9
hub ring, 48
humidity, effect on diskette, 51

IC extractor, 5–6, 17
IC installer, 17
incoming power, checking, 115–116
index hole, on diskette, 47, 48
installation procedures, for components,
    148–168
integrated circuits, 5–6
    damage during installation, 16–17
    tools for, 17–18
internal devices, as source of power supply
        problem, 117–118
internal modems, 162–163
isopropyl alcohol, as drive head cleaner, 142

jewelry, as hazard, 14, 18–19
jumper block, modifying, 150–151

keyboard, 124–127
    checked in self-test, 33
    screen displays for test, 40
keyboard cable, 127
    removing, 129–130
    test for continuity, 131
keyboard plug, 127
keyboard port, 59, 100
    pin locations, 128
knife, as optional tool, 7

LED, on drive, 69
labels for diskettes, 48
    with felt-tip pen, 52
letter-quality printers, 157
light, on drive, 69
line protecting device, 145
liner, as protection for diskette, 51
liquids, as hazard, 26
loading difficulties, with software, 56–57
Log Utilities option, 41

magnetism, effect on diskettes, 53–54
mail order house, support from, 169–170
malfunctions
    error codes for. *See* Error codes
    isolating cause of, 23–26, 29
    quick-reference table for locating source
        of, 178–179
manual, for software, 44
manufacturers, contact with end user, 171
memory, 102–107
    adding to, 104
    amounts on system board, 97
    check with diagnostics diskette, 95
    check with power-on self test, 95
    diagnostics, 104–107
    expansion card for additional, 97
    location on system board, 103
    size and software failure, 56
    switch settings on system board, 193–195
Mode command, 166
modems, 162–164
    not checked by diagnostics, 35
monitors, 133–137
    checked in self-test, 33
    diagnostics for blank, 136–137
    hazards of, 11
    switch settings, 191
    testing cable for continuity, 135
mother board, 95. *See also* System board
multifunction boards, installation of, 164–166
multimeter, 5, 11, 12
    to check power in socket, 32
    proper range setting, 13
    reading while testing drives, 69
    setting to test system board, 97–99, 100
Mylar, 45

needlenose pliers, 6
networking, 166–167
non-IBM devices, and error indication, 108
notes
    of error codes, 41, 107
    importance in diagnosis, 33
    as part of repair process, 21
nut drivers, 6

Ohm's law, 18
one-hand rule, 14
operator. *See* User
options, and false readings during
        diagnostics, 111

Williams: Repair & Maintenance of Your IBM PC (Chilton)

POST. *See* Power-on self-test
paper jam, in printers, 130
parallel drive, pin locations, 189
parallel ports, 100
   maximum, 166
   for printer, 128–129
parallel printer, 158
   adaptor output, 189
Parity Check error, 57
peripherals, moving parts as hazard, 14
Personal Computer Basic display, 34
physical damage
   to computer, 15–17
   to diskette, 51–52
polarity, of components, 21
polyethylene terephthalate, 45
power
   checking for, 32
   to disk drives, 69, 70, 71–73, 121
   on or off, 20
power supply, 111–123
   audio code for malfunction, 34
   automatic shutdown, 116
   connectors and outputs, 114–115
   effect of external devices on, 116–117
   in expansion chassis, 160
   hazards of, 10
   and outputs, 114, 187
   replacement vs repair, 11
   specifications, 113
   voltage tolerances of, 32
power supply cable, for drive, 154, 155
power surges, preventing damage from, 145
power switch, 31
   hazards of, 10
   off during component installation, 149
power-on self-test, 33–34
   error codes, 34, 35, 180
   on keyboard, 126
   normal response of, 34
   for power supply, 116
   system board check, 97
pressure, impact on diskette, 52
preventive maintenance, 26, 139–147
   for printer, 132
   schedule for, 146
price estimate, for repair, 174
printers, 127–133
   cleaning, 132, 140
   diagnostics, 132–133
   DIP switches on, 158–159

installing, 157–159
logging errors to, 41
programs. *See* Software
pulley, on disk drive, 79, 83

quad density drive, 50
quick-reference table, for locating source of
   malfunctions, 178–179

RAM memory, 102
   checked in self-test, 33
   on multifunction board, 165
RAM module
   addresses on expansion board, 107
   addresses on system board, 105, 106, 190
   error codes, 105
   testing, 106–107, 108
ROM (read only memory), 103
read/write, program to test for, 85
read/write head. *See* Disk drive heads
repair contract, 2
repairs
   average down time for, 1
   cost of, 1
   first steps, 3
   replacement as method of, 20–21
   solving problems of, 175
   terms for, 174–175
resistance
   measurement of, 5
   relation with voltage and amperage, 18
   of system board circuitry, 98–99, 100,
     122–123, 185
ribbon, as source of printer problems, 130

safety
   for computer, 15–20
   personal, 8–14
   rules of, 13–14
screen displays
   diagnostics diskette, 37–38
   for keyboard test, 40
screwdrivers, 4
screws, and short circuit hazard, 19
self test. *See* Power-on self-test
serial communications drive, pin locations,
   188
serial ports, 100, 129
   maximum, 164, 166
   outputs, 189
serial printer, 158

track width, on diskette, 47
trade-in value, of components, 21
training, as part of dealer purchase price,
    170

unplugging computer, as safety measure, 13
users. *See also* Operator
    contact with manufacturer, 171
    error as cause of malfunction, 23, 24, 29–
        30
    interaction with technician, 172–174
    responsibilities to dealer, 171–172

Verbatim Corp., disk drive analyzer
    program, 84, 85, 86, 144
visual check, as part of repair process, 24
voltage
    check for keyboard problems, 127
    inside computer, 9
    for disk drive circuit board test, 73

for disk drive power test, 71
in drive as diskette is inserted, 75
limits of power supply, 116
measurement of, 5, 12–13
in monitor, 11
PC tolerance of variations, 32
relation with amperage and resistance, 18
to system board, 100, 119–120, 182
for test of write-protect switch, 79
voltmeter. *See* Multimeter

wall outlet, hazards of, 9–10
warranty, for repair work, 174
weight, impact on diskette, 52
Winchester drives. *See* Hard drives
wire cutter, 7
write-protect notch, 48, 49
write-protect switch
    replacing, 79
    testing, 78–79